# Applied Mathematical Sciences



## EDITORIAL STATEMENT

The mathematization of all sciences, the fading of traditional scientific boundaries, the impact of computer technology, the growing importance of mathematical-computer modelling and the necessity of scientific planning all create the need both in education and research for books that are introductory to and abreast of these developments.

The purpose of this series is to provide such books, suitable for the user of mathematics, the mathematician interested in applications, and the student scientist. In particular, this series will provide an outlet for material less formally presented and more anticipatory of needs than finished texts or monographs, yet of immediate interest because of the novelty of its treatment of an application or of mathematics being applied or lying close to applications.

The aim of the series is, through rapid publication in an attractive but inexpensive format, to make material of current interest widely accessible. This implies the absence of excessive generality and abstraction, and unrealistic idealization, but with quality of exposition as a goal.

Many of the books will originate out of and will stimulate the development of new undergraduate and graduate courses in the applications of mathematics. Some of the books will present introductions to new areas of research, new applications and act as signposts for new directions in the mathematical sciences. This series will often serve as an intermediate stage of the publication of material which, through exposure here, will be further developed and refined. These will appear in conventional format and in hard cover.

### MANUSCRIPTS

The Editors welcome all inquiries regarding the submission of manuscripts for the series. Final preparation of all manuscripts will take place in the editorial offices of the series in the Division of Applied Mathematics, Brown University, Providence, Rhode Island.


SPRINGER-VERLAG NEW YORK INC., 175 Fifth Avenue, New York, N. Y. 10010

*Printed in U.S.A.*

Applied Mathematical Sciences | Volume 29

B. M. Fraeijs de Veubeke

# A Course in Elasticity

Translated by F. A. Ficken
With the Editorial Assistance of D. A. Simons

Springer-Verlag
New York Heidelberg Berlin

B. M. Fraeijs de Veubeke
*formerly:*
Professor of Applied Mechanics
and Aerospace Engineering
University of Liege and Louvain
Belgium

Dr. F. A. Ficken (Translator)
*formerly:*
Professor of Mathematics
New York University
USA

Dr. D. A. Simons (Translation editor)
Division of Engineering
Brown University
Providence, Rhode Island 02912
USA

AMS Subject Classifications: 73-01, 73Cxx

**Library of Congress Cataloging in Publication Data**

Fraeijs de Veubeke, B
A course in elasticity.

(Applied mathematical sciences; v. 29)
Translation of Cours d'élasticité.
1. Elasticity. 2. Elastic analysis (theory of structures) I. Title. II. Series.
QA1.A647 vol. 29 [QA931] 510'.8s
[531'.3823'0151] 79-13318



Printed in the United States of America.

9 8 7 6 5 4 3 2 1

ISBN 0-387-90428-X Springer-Verlag New York Heidelberg Berlin
ISBN 3-540-90428-X Springer-Verlag Berlin Heidelberg New York

# Translation Editor's Preface

This book is based on lecture notes of the late Professor de Veubeke. The subject is presented at a level suitable for graduate students in engineering, physics, or mathematics. Some exposure to linear algebra, complex analysis, variational calculus, or basic continuum mechanics would be helpful.

The first third of the book contains the fundamentals of the theory of elasticity. Kinematics of continuous media, the notions of stress and equilibrium, conservation of energy, and the elastic constitutive law are each treated first in a nonlinear context, then specialized to the linear case.

The remainder of the book is given to three classic applications of the theory, each supplemented by original results based on the use of complex variables. Each one of the three topics - Saint-Venant's theory of prismatic beams, plane deformations, and the bending of plates - is first presented and analyzed in general, then rounded out with numerous specific and sometimes novel examples.

The following notational conventions are generally in force, except where noted to the contrary: lower case boldface letters denote vectors or triples of Cartesian coordinates, upper case boldface letters denote $3 \times 3$ matrices, repeated lower case Latin subscripts are summed over (1,2,3), and non-repeated lower case Latin subscripts are assumed to range over (1,2,3).

The translation editor would like to express his sincere appreciation to Kate MacDougall, both for her superb skill in typing the manuscript and for her cheerful approach to the task. The figures were drawn by Mike Waldygo, whose cooperation is also gratefully acknowledged.

D. A. Simons
Providence, Rhode Island
March 19, 1979

# Contents

# Chapter 1
# Kinematics of Continuous Media

## 1.1. Material and Spatial Coordinates

In a fixed Cartesian frame let $x_i$ denote the coordinates of a material point in a configuration called a reference or initial configuration of an elastic body. (Non-repeated Latin subscripts will be assumed to range over the values 1, 2, 3.) Let $y_i$ denote the coordinates of the same point in a deformed or final configuration. The displacement vector has components $u_i = y_i - x_i$.

One may imagine the displacements and, therefore, the final coordinates expressed as functions of the initial coordinates:

$$u_i = u_i(\mathbf{x}), \quad y_i = x_i + u_i(\mathbf{x}) = y_i(\mathbf{x}). \tag{1.1}$$

(Lower case boldfaced letters will denote vectors, column matrices, or triples of coordinates, depending on the context.) The reference configuration is assumed to be known exactly, and each triplet $\mathbf{x}$ of initial coordinates identifies a well-determined material point. These coordinates are therefore also called *material* coordinates.

The final position of a material point depends on the deformation of the elastic body; it is not, in general, known a priori. Moreover, it is necessary to consider the properties of several final configurations in order to identify which are effectively realized. A triplet $y$ of final coordinates therefore identifies only a point which can be occupied by material points differing according to the configurations analyzed; final coordinates are therefore also called *spatial* coordinates.

Material coordinates are those usually chosen as fundamental variables in elasticity, owing to the fact that one studies the final configuration and its properties through the behavior of a fixed set of material points. The necessary integrations can be performed upon the fixed reference configuration, which usually has simple geometric properties.

In fluid mechanics the material coordinates are generally called *Lagrangian* coordinates. In their general definition, they are formed by each triplet of constants of integration of the differential equations of the trajectories of the particles. In a particular definition one uses the Cartesian coordinates $x_0$ of a particle in a reference configuration; they may be taken, for example, at a conventional state $t_0$ of evolution. This point of view thus agrees with the definition of material coordinates in elasticity.

In most problems of fluid mechanics, however, the engineer is especially interested in an occurrence within boundaries fixed in space, eventually crossed by the particles of the fluid. This spatial configuration is fixed, while the set of particles varies. Fluid mechanics is therefore developed mainly by using spatial coordinates, also called *Eulerian* coordinates. Eulerian formalism considers displacements, along with initial coordinates, as unknown functions of the final coordinates:

$$u_i = u_i(y), \qquad x_i = y_i - u_i(y) = x_i(y). \tag{1.2}$$

In using Eulerian coordinates it is difficult to formulate the constitutive equations in such a manner that relations between stresses and deformations in elasticity can take account of changes of orientation of the preferential directions of anisotropic media. There is no trouble with isotropic media, for which the constitutive equations are invariant with respect to changes in orientation of Cartesian frames. For most fluids one is justified in assuming isotropic properties, but such a reduction for elastic media would be too restrictive. The theory of elasticity will therefore be based essentially on the Lagrangian point of view, as expressed by the equations (1.1).

## 1.2. Neighborhood Transformations

By definition, at a regular point the field $u(x)$ is differentiable. The change of neighborhood

$$dy_i = \frac{\partial y_i}{\partial x_j} dx_j = D_j y_i \, dx_j \tag{1.3}$$

is characterized completely by the elements of the Jacobian matrix

$$J = \{D_j y_i\}, \qquad (D_j = \frac{\partial}{\partial x_j}). \tag{1.4}$$

Formula (1.3) uses, as we shall continue to do, the convention wherein repeated subscripts are summed over (1,2,3). The formula expresses in indicial notation the matrix relation

$$\mathbf{dy} = J\ \mathbf{dx} \tag{1.3'}$$

where $\mathbf{dy}$ and $\mathbf{dx}$ here represent column matrices of Cartesian components. The rules of matrix multiplication thus require that in (1.4) the index $i$ be that of the rows of $J$, and $j$ that of the columns.

The Jacobian, or determinant of the Jacobian matrix, measures the ratio between an element of initial volume $dV$ and the element $d\Omega$ of final volume containing the same material points:

$$d\Omega = (\det J)\ dV. \tag{1.5}$$

Let us imagine now a continuous transformation of the initial configuration toward the final. The value of the Jacobian, initially unity everywhere, varies continuously along the path of each material point. Its value cannot vanish without annulling the volume of a non-empty set of material points; this is physically impossible. The Jacobian can never change sign and, whatever the final configuration, we see that

$$\det J > 0. \tag{1.6}$$

Among other consequences important for the measure of finite deformations, this property assures the existence and uniqueness, at each regular point, of the inverse neighborhood transformation

$$dx_j = \partial_m x_j\ dy_m, \quad (\partial_m = \frac{\partial}{\partial y_m}). \tag{1.7}$$

The easily proven relations

$$\partial_m x_j D_j y_i = \delta_{mi}, \quad D_j y_i \partial_i x_k = \delta_{jk} \tag{1.8}$$

are the expressions in indicial notation of the inversion

relations

$$JJ^{-1} = J^{-1}J = I, \quad J^{-1} = \{\partial_m x_j\}. \tag{1.8'}$$

(Here $\delta_{ij}$ is Kronecker's delta: if $i = j$, $\delta_{ij} = 1$; if $i \neq j$, $\delta_{ij} = 0$.) If the displacement field (1.1) is continuously differentiable on the domain occupied by the initial configuration, then the relation (1.6), valid everywhere, allows application of the theorem on implicit functions. This assures us of the existence and uniqueness of the inverse transformation (1.2).

At a regular point the elements of the Jacobian matrix have a geometric interpretation connected with the natural curvilinear coordinates formed by the Cartesian planes $x_i$ = const. of the initial configuration when they become curved surfaces of the final configuration. Considering

$$\mathbf{y} = y_i(\mathbf{x})\mathbf{e}_i,$$

the final position vector of the material point, the vectors of the local frame of this natural system of curvilinear coordinates are defined by

$$\mathbf{g}_j = D_j\mathbf{y} = D_j y_i \mathbf{e}_i. \tag{1.9}$$

Thus each column of the Jacobian matrix is formed by the Cartesian components of one of the vectors of the local frame; property (1.6) assures the linear independence of these vectors.

The study of properties of the Jacobian matrix is closely related to the study of the properties of the displacement gradient matrix

$$\mathbf{A} = \{D_j u_i\}. \tag{1.10}$$

From (1.1) we find, in fact, the equivalent formulation of a neighborhood transformation

$$dy_i = dx_i + du_i = (\delta_{ji} + D_j u_i) dx_j, \tag{1.11}$$

whence

$$D_j y_i = \delta_{ji} + D_j u_i, \quad \text{or} \quad J = I + A. \tag{1.12}$$

Likewise, from the inverse neighborhood transformation

$$dx_m = dy_m - du_m = (\delta_{im} - \partial_i u_m) dy_i$$

we obtain

$$\partial_i x_m = \delta_{im} - \partial_i u_m, \quad \text{or} \quad J^{-1} = I - X. \tag{1.13}$$

For the inversion relations (1.8') we have the equivalent form

$$XA = AX = A - X. \tag{1.14}$$

## 1.3. Composition of Changes of Configuration

A change of configuration is defined by assigning to each point a displacement $u(r)$ in terms of its initial position vector $r$. Let $u(x)$ be an initial change. It displaces a material point originally at $x$ and assigns it the new position vector $x + u(x)$. This becomes the original position vector of a later displacement $v(r) = v(x + u(x))$. The resulting displacement vector becomes

$$w_{u,v}(x) = u(x) + v(x + u(x)). \tag{1.15}$$

This rule of composition may be compared with the result of reversing the order of changing configuration components:

$$w_{v,u}(x) = v(x) + u(x + v(x)). \tag{1.16}$$

The composite changes (1.15) and (1.16) are generally different, as one could verify by studying, for example, the resultant of rotations about different axes. Consider now neighborhood transformations corresponding to the sequence $u(x)$ followed by $v(r)$. We have

$$dy_i = (\delta_{ji} + D_j u_i)\, dx_j$$

followed by

$$dz_m = (\delta_{im} + \partial_i v_m)\, dy_i$$

resulting finally in

$$dz_m = (\delta_{im} + \partial_i v_m)(\delta_{ji} + D_j u_i)\, dx_j,$$

showing that the composition corresponds to the product of the component Jacobian matrices, taken in the appropriate order. By expanding the product and observing that

$$\partial_i v_m(\delta_{ji} + D_j u_i) = \partial_i v_m D_j y_i = D_j v_m(y) = D_j v_m(x + u),$$

we obtain for the composite Jacobian matrix

$$D_j z_m = \delta_{jm} + D_j[u_m(x) + v_m(x + u)] \tag{1.17}$$

which agrees with (1.15) for the composition of displacements. Now let the field $v(x + u)$ be continuously differentiable and use the theorem of finite increments

$$v(x + u) = v(x) + u_j D_j v(\hat{x}),$$

where

$$\hat{x}_i = x_i + \theta_i u_i \quad \text{(i not summed)}, \quad 0 \leq \theta_i \leq 1.$$

Formula (1.15) becomes

$$w_{u,v}(x) = u(x) + v(x) + u_j D_j v(\hat{x}). \tag{1.15'}$$

We say that the field $v(r)$ satisfies the conditions of *geometric linearity* if the nine gradients $D_j v_m$ satisfy

$$|D_j v_m| \ll 1.$$

Then the displacements represented by the last term of (1.15') are negligible compared with $u(x)$ and we have a linearized law of superposition

$$w_{u,v}(x) = u(x) + v(x).$$

The same law results from inverting the order of changes of configuration and assuming that the field $u(r)$ satisfies the conditions of geometric linearity.

If, therefore, the two displacement fields are geometrically linear, their composition reduces to local superposition and hence is independent of the order in which they are applied.

The theorem of finite increments applied to (1.17),

$$D_j z_m = \delta_{jm} + D_j[u_m(x) + v_m(x) + u_r D_r v_m(\hat{x})],$$

requires, in order to reduce the result to

$$D_j z_m = \delta_{jm} + D_j[u_m(x) + v_m(x)],$$

that the hypothesis of geometric linearity be accompanied by requirements of regularity of growth. If the terms $u_r D_r v_m(\hat{x})$ are negligible in comparison with $u_m(x)$, the same conclusion does not necessarily hold for the comparison of their partial derivatives.

## 1.4. Measure of the State of Local Deformation. Green's and Jaumann's Strain

In a neighborhood transformation (1.11), the square of the distance separating two neighboring points in the final configuration is

$$dy_i dy_i = (\delta_{mi} + D_m u_i)(\delta_{ni} + D_n u_i)\, dx_m dx_n.$$

By expanding and subtracting the square of the initial distance we obtain

$$dy_i dy_i - dx_i dx_i = 2\varepsilon_{mn}\, dx_m dx_n \tag{1.18}$$

where

$$\varepsilon_{mn} = \frac{1}{2}(D_m u_n + D_n u_m + D_m u_i D_n u_i) = \varepsilon_{nm}. \tag{1.19}$$

The coefficients of the quadratic form (1.18) may be arranged in a symmetric matrix

$$E = \begin{pmatrix} \varepsilon_{11} & \varepsilon_{12} & \varepsilon_{13} \\ \varepsilon_{12} & \varepsilon_{22} & \varepsilon_{23} \\ \varepsilon_{13} & \varepsilon_{23} & \varepsilon_{33} \end{pmatrix}. \tag{1.20}$$

They characterize the local deformation of the medium. If they are all zero, in fact, the neighborhood transformation preserves the distance between any two neighboring points. Then, as will be proved in the next section, it must represent a rotation of the neighborhood as a rigid body. Conversely, if the transformation is a rotation and preserves the distance between any two neighboring points, then (1.18) requires that each $\varepsilon_{mn}$ be zero.

The matrix (1.20), which may be expressed in terms of the displacement gradient matrix (1.10) by the relation

$$E = \frac{1}{2}(A + A^T + A^T A) = E^T, \tag{1.21}$$

constitutes a Lagrangian measure of the state of deformation which will be called *Green's* strain.

A neighborhood transformation is a *pure deformation* if its Jacobian matrix is symmetric or, equivalently, if its displacement gradient matrix is symmetric. Then

$$A = H = H^T$$

and (1.21) gives

$$E = H + \frac{1}{2} H^2 \tag{1.22}$$

for Green's strain. Clearly, to each matrix H, there corresponds a unique matrix E. If the converse holds then H itself furnishes another exact Lagrangian measure of the state of deformation. In Chapter 4 we will prove generally the bi-unique correspondence between H and E, and will call H *Jaumann's* strain.* The distinction between Green's and Jaumann's strain is important in the case of finite deformations. For the usual case of infinitesimal deformations, the elements of $H^2$ are negligible in comparison with those of H and Jaumann's strain coincides, practically, with Green's.

## 1.5. Rigid-Body Rotations of a Neighborhood

If the strain vanishes in the neighborhood of a material point, then except for the translation of that point, the neighborhood must be transformed by a rotation. The Jacobian matrix is then orthogonal:

$$J = \{\beta_{ij}\}, \quad \beta_{ij}\beta_{ir} = \delta_{jr}. \tag{1.23}$$

This is a condition necessary and sufficient for the preservation of distances in the neighborhood:

$$dy_i dy_i = (\beta_{ij}dx_j)(\beta_{ir}dx_r) = \beta_{ij}\beta_{ir}\, dx_j dx_r$$
$$= \delta_{jr}\, dx_j dx_r = dx_r dx_r.$$

With $\{\beta_{ij}\} = U$, the relation (1.23) is equivalent to

$$U^T U = I,$$

and we get

$$\det U \cdot \det U^T = (\det U)^2 = 1.$$

The ambiguous sign of the determinant is settled by (1.6), which requires

*Editor's note: No such proof is given.

$$\det U = \det\{\beta_{ij}\} = 1. \tag{1.24}$$

We know that this additional property is necessary and sufficient to guarantee that the orthogonal matrix $U$ is a genuine rotation, neither preceded nor followed by a reflection with respect to a plane (which also preserves distances). We recall finally that, if $U^T$ is a right inverse of $U$, then it is also a left inverse:

$$U^T U U^T = U^T \Rightarrow (U^T U - I)U^T = 0 \Rightarrow U^T U = I,$$

or, using indices,

$$\beta_{ij}\beta_{mj} = \delta_{im}. \tag{1.25}$$

The general structure of a rotation matrix appears simply from a geometric property of the displacement field.

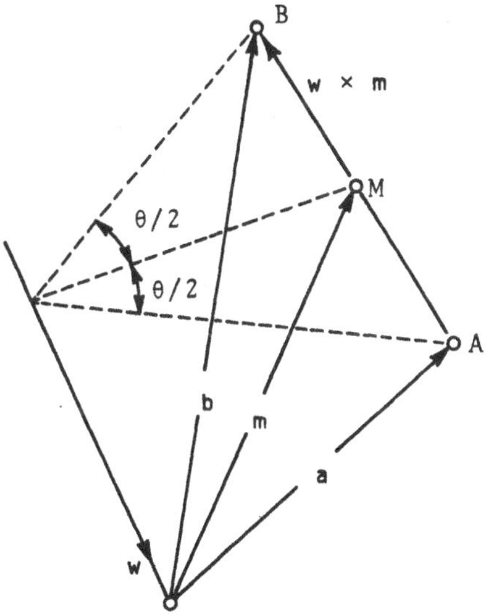


Figure 1.1

If $w$ is a vector along the axis of rotation, its moment $w \times m$ with respect to a point $M$ is perpendicular to the plane containing the axis and passing through $M$. The initial point A and the terminal point B of the moment vector are chosen so that M is the midpoint of the segment AB. If $\mathbf{a}$ and $\mathbf{b}$ are the position vectors of A and B, then

$$\mathbf{m} = \frac{1}{2}(\mathbf{a} + \mathbf{b}), \qquad \mathbf{w} \times \mathbf{m} = \mathbf{b} - \mathbf{a}.$$

The vector $\mathbf{b}$ can be considered as the vector $\mathbf{a}$ rotated about the axis through an angle $\theta$ which is restricted, for the moment, to the interval $0 \leq \theta < \pi$. The modulus of the displacement $\mathbf{b} - \mathbf{a}$ is therefore

$$||\mathbf{w} \times \mathbf{m}|| = 2||\mathbf{m}|| \sin \phi \tan \frac{\theta}{2}$$

where $||\mathbf{m}|| \sin \phi$ is the projection of $\mathbf{m}$ on the plane perpendicular to the axis of rotation. The modulus of the vector product is also

$$||\mathbf{w} \times \mathbf{m}|| = ||\mathbf{w}|| \cdot ||\mathbf{m}|| \sin \phi.$$

Comparison of these two expressions yields

$$||\mathbf{w}|| = 2 \tan \frac{\theta}{2},$$

which depends only on the angle of rotation. The field of moments of this vector is capable, therefore, of representing the displacement field of a finite rotation, provided we consider the moment to be applied to the mean position of each point. The apparently singular limiting case $\theta = \pi$ will be adjusted later. Both the rotation itself and the vector product may be expressed as matrix products:

$$\mathbf{b} = \mathbf{U}\mathbf{a}, \qquad \mathbf{b} - \mathbf{a} = \mathbf{W}\mathbf{m},$$

where the skew-symmetric matrix $\mathbf{W}$ is formed from the components of $\mathbf{w}$:

$$W = \begin{pmatrix} 0 & -w_3 & w_2 \\ w_3 & 0 & -w_1 \\ -w_2 & w_1 & 0 \end{pmatrix} = -W^T.$$

The components of $w$ are then called *strict components* of $W$. The elimination of $m$ and $b$ yields

$$\frac{1}{2} W(I + U)a = (U - I)a,$$

and, since $a$ is arbitrary,

$$\frac{1}{2} W(I + U) = U - I,$$

or after some rearrangement,

$$(I - \tfrac{1}{2}W)U = I + \frac{1}{2} W. \tag{1.26}$$

To obtain $U$ in terms of $W$ we must invert the matrix

$$M = I - \frac{1}{2} W. \tag{1.27}$$

Its characteristic equation

$$\det(M - \alpha I) = 0$$

is extremely easy to develop; one finds

$$(1-\alpha)^3 + \tfrac{1}{4}(1-\alpha)w^Tw = -\alpha^3 + 3\alpha^2 - \alpha(3 + \tfrac{1}{4}w^Tw) + (1 + \tfrac{1}{4}w^Tw) = 0.$$

The determinant of $M$ is equal to the constant term

$$1 + \tfrac{1}{4}w^Tw = 1 + \tan^2 \frac{\theta}{2} = 1/\cos^2 \frac{\theta}{2}.$$

In order to eliminate the singularity at $\theta = \pi$ we multiply the characteristic equation by $\cos^2 \frac{\theta}{2}$. We now appeal to the Cayley-Hamilton theorem, that a matrix always satisfies its characteristic equation. The result is

$$\cos^2 \tfrac{\theta}{2}\{-M^3 + 3M^2 - (3 + \tan^2 \tfrac{\theta}{2})M\} + I = 0.$$

By multiplying by $M^{-1}$, using (1.27), and regrouping, we obtain

$$M^{-1} = I + \frac{1}{2}W \cos^2 \frac{\theta}{2} + \frac{1}{4}W^2 \cos^2 \frac{\theta}{2}.$$

Now we can find the rotation matrix:

$$U = M^{-1}(I - \frac{1}{2}W + W) = I + M^{-1}W$$

$$= I + W + \frac{1}{2}W^2 \cos^2 \frac{\theta}{2} + \frac{1}{4}W^2 \cos^2 \frac{\theta}{2}.$$

To simplify, we observe that

$$W^2 = -w^T wI + ww^T = -4 \tan^2 \frac{\theta}{2}I + ww^T$$

and that

$$Ww = 0.$$

The result is

$$W^3 = -4 \tan^2 \frac{\theta}{2} W$$

and, after substitution and regrouping,

$$U = I + W \cos^2 \frac{\theta}{2} + \frac{1}{2}W^2 \cos^2 \frac{\theta}{2}.$$

A final modification will yield a formula with an arbitrary angle of rotation. With

$$W = 2 \tan \frac{\theta}{2} N,$$

the skew matrix $N$ has for components the direction cosines of a unit vector along the axis of rotation. We arrive at the formulas

$$U = I + \sin \theta \, N + (1 - \cos \theta)N^2. \tag{1.29}$$

and

$$N^2 = -I + nn^T, \quad n^T n = 1. \tag{1.30}$$

This is one of the possible forms of the Cayley-Klein representation.

We have a special interest in infinitesimal rotations. They arise from terms of the first order in the very small angle $\theta$:

$$U = I + \theta N .$$

This result may be written

$$U = I + \Omega \tag{1.31}$$

where $\Omega$ is the skew matrix

$$\Omega = \begin{pmatrix} 0 & -\omega_3 & \omega_2 \\ \omega_3 & 0 & -\omega_1 \\ -\omega_2 & \omega_1 & 0 \end{pmatrix} = -\Omega^T$$

whose elements are the small, superposable rotations

$$\omega_i = \theta\, n_i \tag{1.32}$$

about the Cartesian axes at the point under consideration.

The relation (1.31) arises because the difference between the infinitesimal rotation and the identity is infinitesimal; in the indicial notation we have

$$D_j y_i = \delta_{ij} + \omega_{ij}, \quad |\omega_{ij}| << 1. \tag{1.33}$$

By the conservation of squared distances,

$$D_j y_i D_m y_i = \delta_{jm}.$$

Now

$$(\delta_{ij} + \omega_{ij})(\delta_{im} + \omega_{im}) = \delta_{jm} + \omega_{jm} + \omega_{mj} + \omega_{ij}\omega_{im} = \delta_{jm}.$$

Neglecting terms of the second order, we arrive at the condition for skew symmetry

$$\omega_{jm} + \omega_{mj} = 0. \tag{1.34}$$

Comparison between (1.31) and (1.33) yields

$$\omega_{12} = -\omega_3 \qquad \omega_{23} = -\omega_1 \qquad \omega_{31} = -\omega_2. \tag{1.35}$$

This may be expressed by either of the formulas

$$\omega_i = -\frac{1}{2} e_{imn}\omega_{mn} \qquad \text{or} \qquad \omega_{mn} = -e_{mnp}\omega_p. \tag{1.36}$$

We are using the permutation symbol $e_{mnp}$, where

$$e_{mnp} = \begin{cases} 1, & \text{if the sequence } \{m,n,p\} \text{ is an even permutation of } \{1,2,3\}, \\ -1 & \text{" " odd " " "}, \\ 0 & \text{otherwise.} \end{cases}$$

Thus $e_{mnp}$ is antisymmetric in all its indices, and the only non-zero values are

$$e_{123} = e_{231} = e_{312} = 1, \quad e_{321} = e_{213} = e_{132} = -1.$$

The infinitesimal displacements caused by an infinitesimal rotation about an axis through the origin correspond to a field of moments

$$\mathbf{u} = \boldsymbol{\omega} \times \mathbf{m},$$

because the vector $\boldsymbol{\omega}$ with components (1.32) is comparable with the vector $\mathbf{w}$, and for the moduli, $\theta$ is comparable with $2 \tan \frac{\theta}{2}$. On the other hand, one may compare $\mathbf{m}$ with the position vector $\mathbf{x}$ of the material point at the location of the mean position vector, from which it differs by an infinitesimal semi-displacement. The displacement field may be formulated analytically as

$$u_1 = \omega_2 x_3 - \omega_3 x_2,$$
$$u_2 = \omega_3 x_1 - \omega_1 x_3,$$
$$u_3 = \omega_1 x_2 - \omega_2 x_1,$$

or

$$u_j = e_{jmn}\omega_m x_n. \tag{1.37}$$

It is analogous to the velocity field of a rotating rigid body in the sense that the latter obeys the same equations but with $\omega$ denoting an angular velocity.

## 1.6. The Kinematical Decomposition of the Jacobian Matrix

A general neighborhood transformation may be accomplished in two steps. The first is a pure deformation

$$dz_i = (\delta_{ij} + h_{ij})\, dx_j, \quad h_{ji} = h_{ij},$$

while the second is a rotation

$$dy_m = \beta_{mi}\, dz_i, \quad \beta_{mi}\beta_{mr} = \delta_{ir}, \quad \det\{\beta_{mi}\} = 1.$$

The general transformation would then be represented by

$$dy_m = \beta_{mi}(\delta_{ij} + h_{ij})\, dx_j,$$

which raises the problem of the existence and uniqueness of the *polar decomposition*

$$D_j y_m = \beta_{mi}(\delta_{ij} + h_{ij}). \tag{1.38}$$

The corresponding matrix form is

$$J = U(I + H) \tag{1.39}$$

By multiplying on the left by the transposed relation

$$J^T = (I + H)U^T, \tag{1.39'}$$

and using the fact that $U^TU = I$, we obtain

$$J^TJ = (I + H)^2.$$

By using (1.12) and simplifying,

$$A + A^T + A^TA = 2H + H^2.$$

In view of (1.21), we recover (1.22), which is thus valid not only for a pure deformation, but for a general neighborhood transformation. In the polar decomposition (1.39) the matrix $H$ is Jaumann's strain.

On the other hand, by taking determinants in (1.39), we find

$$\det J = \det U \cdot \det(I + H) = \det(I + H).$$

Because of the physical requirement (1.6), Jaumann's strain must satisfy

$$\det(I + H) > 0. \tag{1.40}$$

The rotation operator is uniquely defined, therefore, by

$$U = J(I + H)^{-1}. \tag{1.41}$$

Elimination of $H$ between (1.39') and (1.41) yields the relation

$$U^T J = (U^T J)^T = J^T U, \tag{1.42}$$

expressing the symmetry of the product $U^T J$.

Except in genuinely exceptional cases, the polar decomposition does not imply the existence of partial displacement fields which would be associated by their gradients with each of the partial transformations, and whose composition would restore the global displacement field.

Now we study the case when the two neighborhood transformations yielding the general transformation are infinitesimal. Formula (1.38) becomes

$$D_j y_m = (\delta_{mi} + \omega_{mi})(\delta_{ij} + h_{ij})$$

with

$$|\omega_{mi}| \ll 1, \qquad |h_{ij}| \ll 1.$$

Expanding and omitting terms of the second order yields the infinitesimal transformation

$$D_j y_m = \delta_{mj} + h_{mj} + \omega_{mj} \tag{1.44}$$

and, for the displacement gradient matrix,

$$D_j u_m = h_{mj} + \omega_{mj} \tag{1.45}$$

Since $h_{mj}$ is symmetric and $\omega_{mj}$ is skew, we finally have the explicit formulas

$$h_{mj} = \frac{1}{2}(D_j u_m + D_m u_j) = h_{jm}, \tag{1.46}$$

$$\omega_{mj} = \frac{1}{2}(D_j u_m - D_m u_j) = -\omega_{jm} \tag{1.47}$$

Since Jaumann's and Green's strain are identical here, it follows from (1.46) that the infinitesimal character of the deformations and rotations allows us to simplify (1.19) by dropping its nonlinear terms.

Let us rewrite (1.47) using the components of $\omega$:

$$2\omega_i = -e_{imj}\omega_{mj} = -\frac{1}{2} e_{imj} D_j u_m + \frac{1}{2} e_{imj} D_m u_j .$$

Now interchange $m$ and $j$ in the first term and regroup:

$$\omega_i = \frac{1}{2} e_{imj} D_m u_j \tag{1.48}$$

or

$$\begin{aligned} \omega_1 &= \frac{1}{2}(D_2 u_3 - D_3 u_2), \\ \omega_2 &= \frac{1}{2}(D_3 u_1 - D_1 u_3), \\ \omega_3 &= \frac{1}{2}(D_1 u_2 - D_2 u_1). \end{aligned} \tag{1.49}$$

The *conditions of geometric linearity*

$$|D_j u_m| << 1, \tag{1.50}$$

when imposed on the nine displacement gradients, render a neighborhood transformation infinitesimal. These conditions are entirely equivalent to the nine distinct conditions

$$|\varepsilon_{jm}| << 1, \quad |\omega_i| << 1. \tag{1.51}$$

## 1.7. Geometric Interpretation of Infinitesimal Strains

Let us consider two material points, one at $\mathbf{x}$, and another with a slightly greater $x_1$ coordinate. Thus $\mathbf{d}_1\mathbf{x} = \{dx_1, 0, 0\}^T$ with $dx_1 > 0$. In the terminal configuration the relative position vector $\mathbf{d}_1\mathbf{y}$ will have components

$$d_1y_1 = (1 + D_1u_1)dx_1, \quad d_1y_2 = D_1u_2dx_1, \quad d_1y_3 = D_1u_3dx_1.$$

It follows that for $\varepsilon_{11}$, (equal to $h_{11}$ to first order),

$$\varepsilon_{11} = D_1u_1 = \frac{d_1y_1 - dx_1}{dx_1} . \tag{1.52}$$

Except for negligible terms, $d_1y_1$ is the final distance between the points. In an infinitesimal deformation, a component of Green's (or Jaumann's) strain with two equal indices indicates, therefore, a *unit elongation* (elongation per unit initial length) in the direction indicated by the indices.

We study now the relations between the vectors $\mathbf{d}_1\mathbf{x}$, $\mathbf{d}_1\mathbf{y}$ and a new vector $\mathbf{d}_2\mathbf{x} = \{0, dx_2, 0\}^T$, $dx_2 > 0$, with image $\mathbf{d}_2\mathbf{y}$ (under the same transformation) having components

$$d_2y_1 = D_2u_1dx_2, \quad d_2y_2 = (1 + D_2u_2)dx_2, \quad d_2y_3 = D_2u_3dx_2.$$

Initially, the two relative position vectors are orthogonal. The angle $\theta_{12}$ between them in the final configuration can be calculated from the scalar product

$$\mathbf{d}_1\mathbf{y} \cdot \mathbf{d}_2\mathbf{y} = \|\mathbf{d}_1\mathbf{y}\| \, \|\mathbf{d}_2\mathbf{y}\| \cos \theta_{12}.$$

On the left we find

$$[(1+D_1u_1)D_2u_1 + D_1u_2(1+D_2u_2) + D_1u_3D_2u_3]\ dx_1dx_2,$$

which reduces to

$$(D_2u_1 + D_1u_2)dx_1dx_2 = 2\varepsilon_{12}\ dx_1dx_2$$

when products of displacement gradients are neglected. On the right, with higher order terms neglected, the respective moduli are $dx_1$ and $dx_2$, and the principal term is

$$\cos\theta_{12}\ dx_1dx_2.$$

The final result is

$$2\varepsilon_{12} = \cos\theta_{12}.$$

The left member is small, by hypothesis, and $\theta_{12}$ is very close to the initial right angle. We may write

$$\theta_{12} = \frac{\pi}{2} - \gamma_{12};$$

since $\cos(\frac{\pi}{2} - \gamma_{12}) = \sin\gamma_{12} \approx \gamma_{12}$, we have

$$2\varepsilon_{12} = \gamma_{12}. \qquad (1.53)$$

In the Green's (or Jaumann's) strain matrix for an infinitesimal deformation, if we double an off-diagonal element we obtain the small decrease in the initial right angle between the indexed directions; the angle of reduction is called the *engineering shear strain.*

A deeper study of the state of a finite or infinitesimal deformation may be undertaken after learning that the strain matrices E of Green or H of Jaumann contain the components, in a given Cartesian frame, of entities known as tensors, certain properties of which are independent of the frame used.

## 1.8. The Eulerian Viewpoint in Kinematics. Almansi's Strain

The Eulerian measure of a finite deformation may be based, just as the Lagrangian measure, on the change in the distances between the points of a neighborhood. Now we use the inverse neighborhood transformation (1.7) to obtain

$$dx_j dx_j = \partial_m x_j \partial_n x_j \, dy_m dy_n,$$

whence

$$dy_m dy_m - dx_j dx_j = (\delta_{mn} - \partial_m x_j \partial_n x_j) dy_m dy_n = 2\phi_{mn} dy_m dy_n. \quad (1.54)$$

The quadratic form now uses the final neighborhood coordinates. The coefficients

$$\phi_{mn} = \frac{1}{2}(\delta_{mn} - \partial_m x_j \partial_n x_j) = \partial_m u_n + \partial_n u_m - \partial_m u_j \partial_n u_j = \phi_{nm} \quad (1.55)$$

may be arranged in a symmetric matrix $F$ which is designated as *Almansi's* strain. We thus have

$$I - (I - X)^T (I - X) = X + X^T - X^T X = 2F. \quad (1.56)$$

Almansi's strain is in a biunique relation with another, having matrix $K$, corresponding to an effective polar decomposition of the inverse Jacobian matrix

$$I - X = V(I - K) \quad (1.57)$$

where $V$ is a rotation and $K$ is symmetric. In fact

$$(I - X)^T (I - X) = I - 2F = (I - K)\, V^T V (I - K) = (I - K)^2$$

and the two Eulerian strain measures are related by

$$F = K - \frac{1}{2}K^2. \quad (1.58)$$

We observe that the order of the factors in the polar decomposition (1.57) is a physical reversal of that used in (1.39). We invert (1.57), recalling that this requires interchanging the factors:

$$I + A = (I - K)^{-1} V^T .$$

We note that the rotation now precedes the deformation. In fact, from

$$I + A = U(I + H) = [U(I + H)U^T]U$$

we find by comparison

$$V^T = U, \quad (I - K)^{-1} = U(I + H)U^T. \tag{1.59}$$

Though not a consequence of the biunique relation between $K$ and Jaumann's strain $H$, nor of the biunique relation between Almansi's and Green's strain, the appearance of the rotation operator is explained easily by the orientation of the measures. Jaumann's strain, as well as Green's, is that of a local observer whose axes suffer the material rotation, for in the polar decomposition (1.57) the measure is taken before turning the material. The strain $K$ (or Almansi's strain), on the other hand, is seen from fixed Cartesian axes.

## 1.9. Eulerian Measures of Rates of Deformation and Rotation

Here we study the continuous evolution of a configuration depending on a supplementary parameter $t$, which may denote time. Equations (1.1) and (1.2) become

$$u_i = u_i(x;t), \quad y_i = x_i + u_i(x;t) = y_i(x;t), \tag{1.1'}$$

$$u_i = u_i(y;t), \quad x_i = y_i - u_i(y;t) = x_i(y;t). \tag{1.2'}$$

Finite increments may be considered as a succession of infinitesimal increments, or as an integration of rates of change. In the Eulerian viewpoint, the velocities of displaced particles are considered as functions of the *spatial* coordinates. The velocities result, however, from differentiating with respect to time, with the *material* coordinates fixed. We denote the latter operation by $D_t$ and refer to it as the *material* time derivative. Then the components of velocity are given by

$$D_t y_i(x;t) = D_t u_i(x;t) = v_i(x,t). \tag{1.60}$$

To get the Eulerian representation $v_i(y;t)$ for the velocity field we should carry out in (1.60) the change of variables (1.2').

When any quantity $f(y;t)$ is represented in spatial coordinates, its *local* time derivative, with spatial coordinates fixed, will be denoted by $\partial_t$. These two types of time derivatives are related by

$$D_t f = \partial_t f + \frac{\partial f}{\partial y_i} D_t y_i = \partial_t f + v_i \partial_i f. \tag{1.61}$$

In the neighborhood of a particle at $y$ at time $t$, the velocity field consists of the velocity $v$ of the particle and an increment

$$dv_i = \partial_j v_i dy_j. \tag{1.62}$$

It will be helpful to decompose this field into two parts,

$$dv_i = (\Omega_{ij} + \theta_{ij})dy_j, \tag{1.63}$$

where

$$\Omega_{ij} = \frac{1}{2}(\partial_j v_i - \partial_i v_j) = -\Omega_{ji} \tag{1.64}$$

is the skew part of $\partial_j v_i$, and

$$\theta_{ij} = \frac{1}{2}(\partial_j v_i + \partial_i v_j) = \theta_{ji} \tag{1.65}$$

is the symmetric part.

We now calculate the rate of growth of the squared distance between the particle and another chosen arbitrarily in the neighborhood:

$$D_t\{dy_i dy_i\} = 2dy_i D_t dy_i = 2dy_i d(D_t y_i) = 2dy_i dv_i. \qquad (1.66)$$

Because of skewness,

$$\Omega_{ij}\, dy_i dy_j = 0,$$

and substitution of (1.63) in (1.66) yields

$$D_t\{dy_i dy_i\} = 2\theta_{ij}\, dy_i dy_j.$$

By analogy with (1.54) we consider the elements (1.65) of the symmetric matrix $\theta$ as Eulerian measures of the *deformation rate* of the neighborhood. With the skew part we associate the vector with components

$$\Omega_m = -\frac{1}{2}\, e_{mij}\Omega_{ij} = \frac{1}{2}\, e_{mij}\partial_i v_j. \qquad (1.67)$$

The part of $dv_i$ given by

$$\Omega_{ij} dy_j = e_{imj}\Omega_m dy_j \qquad (1.68)$$

expresses the velocity field of the neighborhood rotating as a rigid body. This decomposition, due to Helmholtz, expresses the velocity field in the neighborhood of a particle as a translation with the velocity of the particle, a rigid rotation about the particle with angular velocity vector $\frac{1}{2}$ rot v (called the vorticity vector), and a remainder which represents the *deformation rate* by a symmetric matrix with elements given by (1.65).

If we adopt the configuration at time $t$ as the initial configuration, we may identify $\partial_j$ with $D_j$. If also we adopt the configuration at time $t + dt$ as the final configuration, we recover the results (1.45), (1.46), and (1.47) previously established for an infinitesimal change of configuration. It is enough to identify $u_i$ with $v_i dt$, $h_{mj}$ with $\theta_{mj} dt$, and $\omega_{mj}$ with $\Omega_{mj} dt$.

We see again the complete equivalence of the Lagrangian and Eulerian formulations of the kinematics of infinitesimal deformations and rotations.

## 1.10. Temporal Variation of the Polar Decomposition of the Jacobian Matrix

It is natural to ask how the Helmholtz decomposition may be related to the temporal variation of the decomposition of the Jacobian. By applying $D_t$ to (1.38) we obtain

$$D_t D_j y_m = D_j D_t y_m = D_j v_m = (\delta_{ij} + h_{ij}) D_t \beta_{mi} + \beta_{mi} D_t h_{ij}. \quad (1.69)$$

If we multiply (1.38) by $\beta_{mr}$, there follows

$$\beta_{mr} D_j y_m = \beta_{mr} \beta_{mi} (\delta_{ij} + h_{ij}) = \delta_{ri} (\delta_{ij} + h_{ij}) = \delta_{rj} + h_{rj},$$

and if we multiply this by $\partial_p x_j$, we obtain

$$\beta_{mr} \partial_p x_j D_j y_m = \beta_{mr} \delta_{pm} = \beta_{pr} = \partial_p x_j (\delta_{rj} + h_{rj}).$$

The last two formulas allow the replacement, with appropriate adjustments of indices, of the values of $\delta_{ij} + h_{ij}$ and $\beta_{mi}$ in the right hand side of (1.69), which becomes

$$D_j v_m = D_j y_n \beta_{ni} D_t \beta_{mi} + \partial_m x_p (\delta_{ip} + h_{ip}) D_t h_{ij}.$$

After multiplication by $\partial_q x_j$,

$$\partial_q v_m = \beta_{qi} D_t \beta_{mi} + \partial_q x_j \partial_m x_p (\delta_{ip} + h_{ip}) D_t h_{ij}. \quad (1.70)$$

This formula resembles the Helmholtz decomposition

$$\partial_q v_m = \Omega_{mq} + \theta_{mq}.$$

The first term of the right hand side is skew and has the character of a component of rotation

$$\begin{aligned} W_{mq} &= \beta_{qi} D_t \beta_{mi} = D_t(\beta_{qi}\beta_{mi}) - \beta_{mi} D_t \beta_{qi} \\ &= -\beta_{mi} D_t \beta_{qi} = -W_{qm}. \end{aligned} \tag{1.71}$$

The second term has a symmetric part, but the rest may not be:

$$\partial_q x_j \partial_m x_p (\delta_{ip} + h_{ip}) D_t h_{ij} = \partial_q x_j \partial_m x_p D_t h_{pj} + \partial_q x_j \partial_m x_p h_{ip} D_t h_{ij}.$$

In fact, for the symmetric and skew parts of (1.70)

$$\theta_{mq} = \partial_q x_j \partial_m x_p \{D_t h_{pj} + \tfrac{1}{2} h_{ip} D_t h_{ij} + \tfrac{1}{2} h_{ij} D_t h_{ip}\} \tag{1.72}$$

$$\Omega_{mq} = W_{mq} + \tfrac{1}{2} \partial_q x_j \partial_m x_p \{h_{ip} D_t h_{ij} - h_{ij} D_t h_{ip}\}. \tag{1.73}$$

From (1.73), if $h_{ij} \equiv 0$ at the instant studied, then the component of real rotation will coincide with $W_{mq}$ and the deformation rate will be proportional to the time derivative of Jaumann's strain. In this case

$$\theta_{mq} = D_t h_{pq} \quad \text{and} \quad \Omega_{mq} = W_{mq}$$

if the configuration at the instant studied is chosen as the initial configuration.

By using (1.22), we get a more suggestive form for (1.72):

$$\theta_{mq} = \partial_q x_j \partial_m x_p D_t \varepsilon_{pj}. \tag{1.74}$$

The deformation rate is proportional to the material time derivative of Green's strain, in spite of the fact that Green's measure is not of Eulerian type.

We shall now show that the deformation rate may be related, in a clearly Eulerian way, to Almansi's strain. From the expression for Green's strain

$$2\varepsilon_{ij} = D_j y_m D_i y_m - \delta_{ij}$$

we obtain, using (1.55),

$$2\partial_p x_j \partial_q x_i \varepsilon_{ij} = \delta_{pq} - \partial_p x_j \partial_q x_j = 2\phi_{pq}.$$

We may now adjust (1.74) to get

$$\begin{aligned}\theta_{mq} &= D_t(\partial_q x_j \partial_m x_p \varepsilon_{pj}) - \partial_m x_p \varepsilon_{pj} D_t \partial_q x_j - \partial_q x_j \varepsilon_{pj} D_t \partial_m x_p \\ &= D_t \phi_{mq} - \partial_m x_p \varepsilon_{pj} D_t \partial_q x_j - \partial_q x_j \varepsilon_{pj} D_t \partial_m x_p. \qquad (1.75)\end{aligned}$$

Now

$$\partial_p v_q = \partial_p x_j D_j v_q = \partial_p x_j D_t D_j y_q = D_t(\partial_p x_j D_j y_q) - D_j y_q D_t \partial_p x_j.$$

In the last member the first term vanishes because $\partial_p x_j D_j y_q = \delta_{pq}$, whence, multiplying both sides by $\partial_q x_r$,

$$\partial_q x_r \partial_p v_q = -D_j y_q \partial_q x_r D_t \partial_p x_j = -D_t \partial_p x_r.$$

Using this type of result in (1.75) finally yields

$$\theta_{mq} = D_t \phi_{mq} + \phi_{mr} \partial_q v_r + \phi_{rq} \partial_m v_r. \qquad (1.76)$$

The right hand side represents *Lie's derivative* of Almansi's strain.

# Chapter 2
# Statics and Virtual Work

## 2.1. The Concept of Stress. True Stress

In the interior of a continuous medium we isolate a set of points which is contained, in the final configuration, in a simply connected domain $\Omega$, bounded by a simple surface $\partial\Omega$. Among the exterior forces affecting these points we distinguish:

1. Those which are proportional to mass, such as gravity or inertia;

2. Those arising from the action of other material points in the medium but outside the domain $\Omega$.

The first act at a distance. The others are intermolecular, with a radius of action so short as to affect only the particles very close to the surface $\partial\Omega$. The idealization of this situation, due to *Euler* and *Cauchy*, presents the vector sum of the exterior forces as a volume integral for forces acting at a distance and a surface integral for the others:

$$\int_{\Omega} \rho g \, d\Omega + \int_{\partial\Omega} \tau \, d\Sigma .$$

Here $\rho$ is the density of the medium, $g$ the *body force* vector, and $\tau$ the *surface traction* vector representing intermolecular forces.

The Euler-Cauchy formulation has proven satisfactory for the theoretical treatment of the classical problems of elasticity and fluid mechanics. The addition of a surface layer of *stress-couples* introduces complications which have not been applied enough to allow serious confrontation with experience.

The surface traction vector depends on the orientation of the exterior normal $\nu$ to the surface element $d\Sigma$. Consider first surface elements with exterior normal parallel to one of the Cartesian axes. For an exterior normal in the

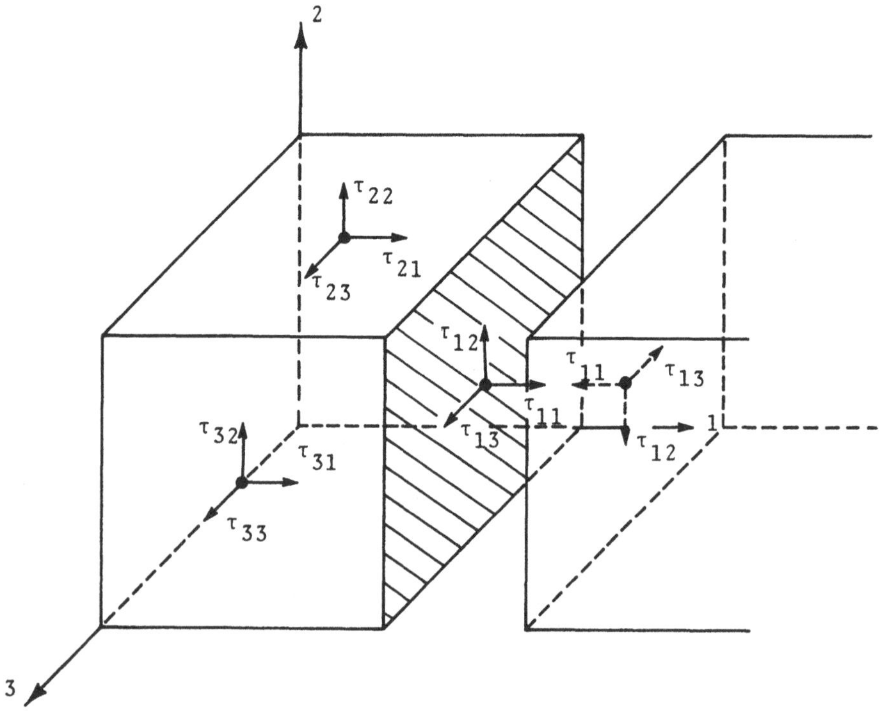


Figure 2.1

direction of the first axis the Cartesian components of $\tau$ are $(\tau_{11},\tau_{12},\tau_{13})$, the first index referring to the normal, and the second to the component. Similar notations $(\tau_{21},\tau_{22},\tau_{23})$ and $(\tau_{31},\tau_{32},\tau_{33})$ are used for surface elements with normals parallel to the second and third axes respectively. As the dimensions of the parallelepiped approach zero, the array

$$\begin{pmatrix} \tau_{11} & \tau_{12} & \tau_{13} \\ \tau_{21} & \tau_{22} & \tau_{23} \\ \tau_{31} & \tau_{32} & \tau_{33} \end{pmatrix} \tag{2.1}$$

of the components of $\tau$ forms, by definition, the matrix of the state of stress.

If the direction of the exterior normal is opposite to that of an axis, the positive direction of stress on that face is also reversed. This convention displays simply the law of action and reaction. If a second element of volume is applied against the shaded face in Figure 2.1, the stresses $(\tau_{11},\tau_{12},\tau_{13})$ represent the effect of the contact by the second on the first. The forces exerted by the first element on the second are expressed by stresses of opposite direction, just as the direction of the exterior normal to the second element is opposite to that of the first. Thus a single collection of algebraic values expresses the reciprocal intermolecular forces on the interface.

Now we consider the relation between the state of stress and the surface traction vector for elements of arbitrary orientation. Figure 2.2 shows an elementary tetrahedron located at an arbitrary interior point 0, with three faces

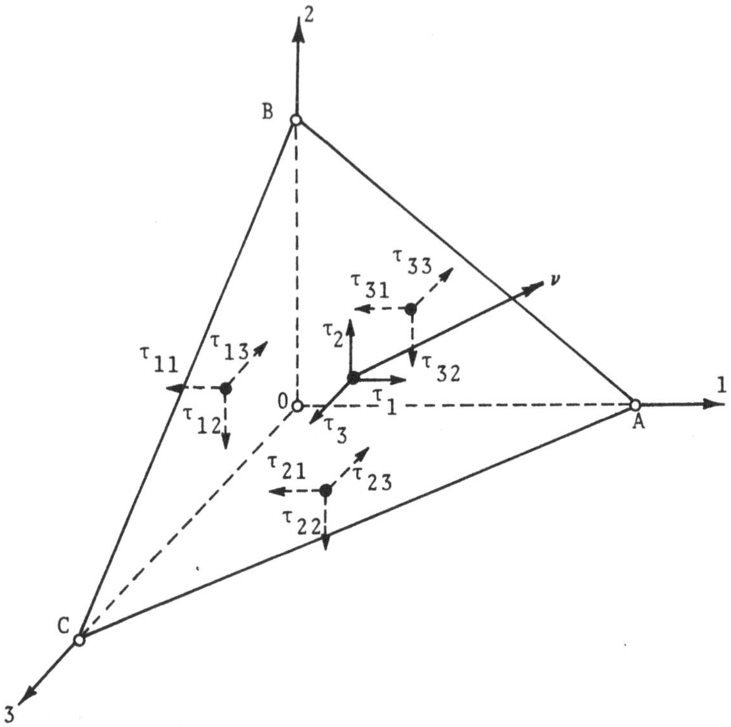


Figure 2.2

parallel to the coordinate planes and the orientation of the fourth specified by its exterior normal $\nu = \nu_i \mathbf{e}_i$. By the Euler-Cauchy principle, the equilibrium of the element under the action of the exterior forces is expressed by

$$\int_\Omega \rho \mathbf{g}\, d\Omega + \int_{OBC} \boldsymbol{\tau}\, d\Sigma + \int_{OCA} \boldsymbol{\tau}\, d\Sigma + \int_{OAB} \boldsymbol{\tau}\, d\Sigma + \int_{ABC} \boldsymbol{\tau}\, d\Sigma = 0.$$

As the dimensions of the tetrahedron approach zero with the orientation of the inclined face fixed, the volume integral approaches zero more rapidly than the surface integral. According to the preceding definitions, the components of $\boldsymbol{\tau}$ on OBC may be written $(-\tau_{11}, -\tau_{12}, -\tau_{13})$. In the same way, we have $(-\tau_{21},-\tau_{22},-\tau_{23})$ on OCA and $(-\tau_{31},-\tau_{32},-\tau_{33})$ on OAB. With $(\tau_1,\tau_2,\tau_3)$ denoting the components of $\tau$ on ABC, for equilibrium we require

$$-\tau_{ij}\, d\Sigma_i + \tau_j\, d\Sigma = 0,$$

where $d\Sigma_i$ is the surface element with normal opposite to the $x_i$-axis. Since by geometry $d\Sigma_i = \nu_i\, d\Sigma$, we obtain *Cauchy's formula*

$$\tau_j = \nu_i \tau_{ij}. \tag{2.2}$$

This gives the components of surface traction explicitly in terms of the stress and the direction cosines of the exterior normal to the respective face. According to the law of action and reaction, the components of surface traction change sign with a reversal of the direction of the normal.

## 2.2. The Piola Stresses

The state of stress depends essentially on the distortion of the molecular array and is therefore a function of the final configuration. In other words, this concept is fundamentally Eulerian. The stresses which have been defined by isolating simple geometric elements in the final configuration are the *true* or *Eulerian stresses*, and are the ones used in fluid mechanics. By contrast, in the theory of elasticity it is helpful to use Lagrangian or material variables. The definition of the state of stress can be modified by relating it to the initial configuration. The elementary tetrahedron is now located there, and the forces $dF_i$ and $dF$ appearing on its faces *after deformation* are related to initial areas $dS_i$ and $dS$. Thus we define new surface tractions

$$t_i = \lim_{dS_i \to 0} \frac{dF_i}{dS_i} \quad \text{and} \quad t = \lim_{dS \to 0} \frac{dF}{dS}$$

with $dS_i = n_i dS$; here $n_i$ are the direction cosines of the exterior normal to the face in the initial configuration. The equilibrium equation

$$dF_1 + dF_2 + dF_3 + dF = 0$$

then takes the form

$$n_1 t_1 + n_2 t_2 + n_3 t_3 + t = 0$$

whence follows the alternate form of Cauchy's formula

$$t_j = n_i t_{ij}, \tag{2.3}$$

where the $t_{ij}$ are the Cartesian components in the $x_j$ direction of the new surface tractions for the faces with exterior normal initially directed along the $x_i$-axis.

The components $t_{ij}$ supply a hybrid representation of the stress, because the forces occurring in the final configuration are manipulated in the geometry of the initial configuration. For this reason some authors call them *Euler-Lagrange* stresses. They are better known as the *Piola* stresses. Their matrix is

$$T = \begin{pmatrix} t_{11} & t_{12} & t_{13} \\ t_{21} & t_{22} & t_{23} \\ t_{31} & t_{32} & t_{33} \end{pmatrix} \tag{2.4}$$

Analysis of the geometry of surface deformations would yield the relation between Piola's stresses and the true stresses, but we shall not pursue this matter.

Besides the above two, other measures of the state of

stress may be defined. For example, the *Kirchhoff-Trefftz* stresses constitute a purely Lagrangian measure. In any problems involving large deformations the distinctions among the various measures must be observed. Each has its own special merits and disadvantages (some of which we shall demonstrate in the sequel): the true stresses are symmetric in their indices but their translational equilibrium equations take a simple form only in spatial coordinates, the Piola stresses satisfy translational equilibrium equations of simple form in material coordinates but are not symmetric, and the Kirchhoff-Trefftz stresses, due to their Lagrangian nature, are convenient for the formulation of constitutive relations but fail to satisfy simple equilibrium equations. When the deformations are infinitesimal, however, all stress measures become practically equivalent.

## 2.3. Translational Equilibrium Equations

The contact force $\mathbf{dF}$ on a surface element, as it actually occurs in the final configuration, has been simply expressed in the geometry of the initial configuration by using Piola's stresses; its Cartesian components are given by

$$dF_j = n_i t_{ij} \; dS. \tag{2.5}$$

For the body forces, one uses the conservation of mass of an element

$$\rho \; d\Omega = \rho_0 \; dV$$

where $\rho_0$ is the local density of the medium in its reference configuration. In view of (1.5), this conservation equation is equivalent to

$$\rho \det \mathrm{J} = \rho_0. \tag{2.6}$$

The translational equilibrium of a body of volume $V$, contained initially in a surface $\partial V$, is expressed by annulling the vector sum of the exterior forces:

$$\int_V \rho_0 g_j \, dV + \int_{\partial V} n_i t_{ij} \, dS = 0.$$

Application of the divergence theorem to the second term yields

$$\int_V (\rho_0 g_j + D_i t_{ij}) dV = 0. \tag{2.7}$$

Since the volume $V$ may be chosen arbitrarily, (2.7) yields the *translational equilibrium equations*

$$D_i t_{ij} + \rho_0 g_j = 0. \tag{2.8}$$

Piola's stresses have the advantage of satisfying *linear* translational equilibrium equations in the material coordinates

## 2.4. Rotational Equilibrium Equations

The rotational equilibrium of a portion of the volume of a continuous medium is expressed by annulling the moment about the origin of the exterior forces acting upon it. It is important to keep in mind that the forces are applied to material points in their final configuration.

For an element of contact force the moment with respect to the origin is

$$(\mathbf{x} + \mathbf{u}) \times d\mathbf{F}$$

with the $m^{th}$ component

$$e_{mnj}(x_n + u_n) dF_j .$$

For the volume $V$ the equations of rotational equilibrium are

$$\int_V e_{mnj}\rho_0(x_n+u_n)g_j\, dV + \int_{\partial V} e_{mnj}(x_n+u_n)n_i t_{ij}\, dS = 0.$$

We again apply the divergence theorem, which yields

$$\int_V e_{mnj}\{\rho_0(x_n+u_n)g_j + D_i[(x_n+u_n)t_{ij}]\}dV = 0.$$

Again the volume is arbitrary, so we find

$$e_{mnj}\{\rho_0(x_n+u_n)g_j + D_i[(x_n+u_n)t_{ij}]\} = 0.$$

Using the simplification afforded by (2.8), we obtain

$$e_{mnj}t_{ij}D_i(x_n + u_n) = 0. \tag{2.9}$$

This implies that the expression

$$t_{ij}(\delta_{in} + D_iu_n) = t_{in}(\delta_{ij} + D_iu_j) \tag{2.10}$$

multiplying the permutation symbol must be symmetric in the indices $n$ and $j$. The relation is trivial for $n = j$. Therefore, there are only three non-trivial equations for rotational equilibrium, and they arise when $j \neq n$. In contrast with the translational equations, they are essentially nonlinear, for the displacements are unknowns as well as the stresses. In contrast with the matrices of the various strain measures, the matrix of Piola's stresses is not symmetric. In Section 2.6 we shall see that it is fundamentally conjugate to the displacement gradient matrix, which is also nonsymmetric.

## 2.5. Statics and Virtual Work

The designations "translational" and "rotational equilibrium equations" suggest that these equations could be derived by *analytical statics*, i.e., by application to statics of the principle of virtual work. In this section we shall show that a form of the principle pertaining to rigid bodies does in fact yield the equilibrium equations.

Let a *virtual displacement* $\delta u$ be defined as any vector field satisfying the internal kinematic constraints of a medium. Thus for a rigid medium it must represent a rigid body motion, while for a deformable medium it need only be differentiable. A virtual displacement may also be called a *variation* and regarded as the difference

$$\delta u(x) = \hat{u}(x) - u(x) \tag{2.11}$$

between the actual field $u$ and any other field $\hat{u}$ consistent with the internal constraints.

If $g_j$ and $t_{ij}$ are the components of body force and Piola's stress, the *virtual work* $\delta$ of the exterior forces on a region $V$ of a continuous medium is defined as

$$\delta = \int_V \rho_o g_j \delta u_j \, dV + \int_{\partial V} n_i t_{ij} \delta u_j \, dS. \tag{2.12}$$

For *rigid bodies* the principle of virtual work states that a necessary and sufficient condition for equilibrium is that the virtual work *vanish* for all virtual displacements.

The virtual displacements of rigid bodies are constrained to preserve distances between points. Suppose that for a deformable solid the virtual work vanishes for all rigid virtual displacements. By the divergence theorem applied to (2.12),

$$\delta = \int_V \{\rho_0 g_j \delta u_j + D_i(t_{ij}\delta u_j)\} dV = 0. \tag{2.13}$$

For virtual translations

$$\delta u_j = dc_j = \text{const.},$$

$D_i \delta u_j = 0$ and (2.13) becomes

$$dc_j \int_V (\rho_0 g_j + D_i t_{ij}) dV = 0.$$

Since the constants are arbitrary each integral vanishes, and because the region $V$ is arbitrary the integrands vanish point-by-point. Thus there follow the translational equilibrium equations (2.8). Infinitesimal virtual rigid rotations about the origin take the form

$$\delta u_j = e_{jmn} d\omega_m (x_n + u_n),$$

where $d\omega_m$ are components of the rotation vector, and convert (2.13) to the form

$$e_{jmn} d\omega_n \int_V [(\rho_0 g_j + D_i t_{ij})(x_n + u_n) + t_{ij}(\delta_{in} + D_i u_n)] dV = 0.$$

In view of the translational equilibrium equations and the arbitrariness of $d\omega_n$ this reduces to

$$\int_V e_{jmn} t_{ij} (\delta_{in} + D_i u_n) dV = 0.$$

Because $V$ is arbitrary, the integrand vanishes, and there follow the rotational equilibrium equations (2.9) and their consequential symmetry relation (2.10).

By reversing the steps, one easily proves the converse, so the vanishing of the virtual work for all rigid virtual displacements is a necessary and sufficient condition for the differential equations of equilibrium (2.8) and (2.9) to hold.

## 2.6. Commutativity of the Operators $\delta$ and $D_i$

The operator $\delta$ forms, for the same values of material coordinates, the difference between two nearly equal fields of intensive quantities. Such differences may be called material or Lagrangian variations. Equations (2.11) are an example for the displacement field. When a family of configurations depends continuously upon a parameter, as for example in equations (1.1') (cf. p. 23), an associated material variation may be defined in the form

$$\delta u_j = dt\, D_t u_j(x,t) = dt\, v_j. \tag{2.15}$$

In this relation, if $t$ is the time in a real evolution of the medium, the variation $\delta u_j$ is a real variation. It is clear that if $\delta u(x,t)$ satisfies the conditions for interchanging the order of partial derivatives, then

$$D_i \delta u_j = dt\, D_i D_t u_j(x,t) = dt\, D_t D_i u_j = \delta D_i u_j. \tag{2.16}$$

More generally, the parameter $t$ could represent a virtual evolution of the medium from a final configuration characterized by the displacements $u_j(x,0)$. In this case the result of the preceding commutativity prevails, but the virtual variation from the final configuration is obtained by setting $t = 0$ after performing the differentiations. Also, the quantity $v_j$ in (2.15) is a *virtual velocity*. Thus it is clear that all conclusions reached by using virtual displacements could have been based on virtual velocities.

If we now multiply (1.61) by an infinitesimal increment of the parameter $t$, we obtain

$$dt\, D_t f = dt\, \partial_t f + v_i\, dt\, \partial_i f$$

which we may interpret as

$$\delta f = dt\, \partial_t f + \delta u_i \partial_i f. \tag{2.17}$$

This formula relates the material variation (real or virtual) of the intensive quantity $f(y,t)$ with its Eulerian or spatial variation $dt\, \partial_t f$, characterizing the perturbation of its field at a fixed point in space. Let us compare the spatial derivative of (2.17),

$$\partial_j \delta f = dt\, \partial_j \partial_t f + \delta u_i \partial_j \partial_i f + (\partial_i f) \partial_j \delta u_i,$$

with the result of applying the operator $\delta$ of (2.17) to $\partial_j f$ in place of $f$ itself,

$$\delta \partial_j f = dt\, \partial_t \partial_j f + \delta u_i \partial_i \partial_j f.$$

We see that, if the order of differentiation of $f(y;t)$ may be reversed,

$$\partial_j \delta f = \delta \partial_j f + (\partial_i f) \partial_j \delta u_i. \tag{2.18}$$

Thus while the operator $\delta$ of material variation commutes with the operators $D_i$ of partial differentiation with respect to material coordinates, it does not commute with the operators $\partial_j$ of differentiation with respect to the spatial coordinates.

There is commutativity, however, between local temporal variations and spatial differentiation:

$$dt\ \partial_t(\partial_j f) = \partial_j(dt\ \partial_t f). \tag{2.19}$$

## 2.7. Virtual Work in a Continuous Medium

Reconsider the general expression (2.13) of virtual work for arbitrary $\delta u$:

$$\delta = \int_V [(\rho_o g_i + D_i t_{ij})\delta u_j + t_{ij} D_i \delta u_j] dV.$$

This may be simplified by the equations (2.8) for translational equilibrium. With the commutativity of $\delta$ and $D_i$ it becomes

$$\delta = \int_V t_{ij} \delta a_{ji}\ dV \tag{2.20}$$

where we have written

$$a_{ji} = D_i u_j \tag{2.21}$$

for the elements of the displacement gradient matrix. In terms of Piola's stresses, the principle of virtual work for a continuous medium is that the equality of (2.12) and (2.20) is necessary and sufficient for equilibrium.

The integrand of (2.20) is the density of virtual work, whose value is here referred to volume in the initial configuration. It is more generally represented by a scalar product of conjugate matrices, one representing the state of stress, the other the state of deformation.

In the present case this product is

$$\mathrm{tr}(\mathbf{T}\delta\mathbf{A}) \tag{2.22}$$

where neither $\mathbf{A}$ nor $\mathbf{T}$ is symmetric. Since the trace of a product of two matrices is invariant under commutation of factors as well as under transposition of the product, we have the equivalent expressions:

$$\mathrm{tr}(T\delta A) = \mathrm{tr}(\delta AT) = \mathrm{tr}(T^T\delta A^T) = \mathrm{tr}(\delta A^T T^T).$$

One sees from its origin that the principle of virtual work involves

1. a kinematical condition: the compatibility of the virtual displacement $\delta u$ and the virtual displacement gradient $\delta A$ as expressed by equation (2.21), and

2. a statical condition: the translational and rotational equilibrium of stresses and body forces.

On the other hand, the principle does not invoke any particular formulation of constitutive equations relating the state of deformation to the stress and describing the physical nature of the medium (e.g., elastic solid, fluid, or more general rheological medium); it may therefore be applied to all continuous media.

## 2.8. Statics and Virtual Power for True Stresses

Our remarks on the coalescence of the various definitions of the state of stress, when the rotations and material deformations are infinitesimal, prompt us to establish the exact structure of the equilibrium equations satisfied by the true stresses. This will be a simple exercise in analytical statics. It is somewhat more difficult to establish the exact nature of the deformation measure which is conjugate, in the sense of virtual work, to the true stresses.

As a consequence of the very definition of true stresses, the calculations are based here on the final configuration. The virtual work is

$$\delta = \int_\Omega \rho g_j \delta u_j \, d\Omega + \int_{\partial\Omega} \nu_i \tau_{ij} \delta u_j \, d\Sigma,$$

or, after applying the divergence theorem,

$$\delta = \int_\Omega [\rho g_j \delta u_j + \partial_i(\tau_{ij}\delta u_j)]d\Omega. \tag{2.24}$$

The vanishing of virtual work for arbitrary virtual translations yields the translational equilibrium equations

$$\rho g_j + \partial_i \tau_{ij} = 0. \tag{2.25}$$

They are again linear but for the use of spatial coordinates; they are the equilibrium equations of fluid mechanics. They allow (2.24) to be reduced to

$$\delta = \int_\Omega \tau_{ij}\partial_i \delta u_j \ d\Omega. \tag{2.26}$$

The rotational displacement variation (2.14) may be rewritten

$$\delta u_j = e_{jmn} \ d\omega_m \xi_n,$$

leading to the very simple result

$$\partial_i \delta u_j = e_{jmn} d\omega_m \delta_{ni} = e_{jmi} \ d\omega_m.$$

Applying this to (2.26) with $\delta = 0$ yields

$$\int_\Omega e_{jmi}\tau_{ij} d\Omega = 0.$$

For an infinitesimal volume, this result simply expresses the symmetry of the matrix of true stresses, i.e.,

$$\tau_{ij} = \tau_{ji}. \tag{2.27}$$

This symmetry permits an adjustment in the expression for virtual work. We interchange the summation indices i and j, then invoke the symmetry of the true stresses:

$$\tau_{ij}\partial_i \delta u_j = \tau_{ji}\partial_j \delta u_i = \tau_{ij}\partial_j \delta u_i.$$

The virtual work now takes the form

$$\delta = \int_\Omega \tau_{ij} \tfrac{1}{2}(\partial_i \delta u_j + \partial_j \delta u_i) d\Omega, \tag{2.28}$$

where the multiplier of $\tau_{ij}$ is now also symmetric. To see the relationship of this factor to the matrix (1.65) of deformation rates we interpret the displacement variations in terms of virtual velocities $\delta u_j = v_j\ dt$, with the result

$$\tfrac{1}{2}(\partial_i \delta u_j + \partial_j \delta u_i) = \theta_{ij}\ dt. \tag{2.29}$$

A first interpretation of the theorem of virtual work is supplied by the equivalent expression of virtual power

$$P = \int_\Omega \tau_{ij}\theta_{ij}\ d\Omega \tag{2.30}$$

where the deformation rates are conjugate to the true stresses to supply the Eulerian density (per unit spatial volume) of virtual power, a quantity which plays an important role in fluid mechanics.

An incremental form more useful in the theory of elasticity uses (2.28) in the form

$$\delta = \int_\Omega \tau_{ij}\Delta\phi_{ij}\ d\Omega \tag{2.31}$$

where, using (2.29), one introduces a *Lie variation* of Almansi's strain

$$\Delta\phi_{ij} = \delta\phi_{ij} + \phi_{ir}\partial_j\delta u_r + \phi_{rj}\partial_i\delta u_r. \tag{2.32}$$

Like Lie's temporal derivative, this operator is perfectly Eulerian and allows Almansi's strain to be regarded as conjugate to the true stress. The basic reason for choosing Lie's variation concerns the idea of an objective differential operator, which arises in connection with spatial coordinates. While waiting to clarify that idea until Chapter 4, we may say that it is justified by the need to calculate rates of growth which are not distorted by a local relative rotation between matter and the frame of the observer.*

## 2.9. Statics and Virtual Work in Infinitesimal Changes of Configuration

The hypothesis $|D_i u_j| << 1$ which characterizes infinitesimal changes of configuration allows (2.10) to be simplified as

$$t_{nj} = t_{jn}, \tag{2.33}$$

and shows Piola's stresses to be symmetric in the first approximation, a natural result since in this instance they should also approximate the true stresses. We notice that (2.33) results directly from neglecting the displacement $u$ compared with $x$ in calculating the moments of forces. This indicates an essential result of the accepted approximation: the equilibrium of the medium and each of its parts is related to the reference configuration instead of the final configuration. We have seen that the same hypothesis of geometric linearization allows the strain to be measured by the *linearized* or *infinitesimal strains*

$$\varepsilon_{ij} = \frac{1}{2}(D_i u_j + D_j u_i). \tag{2.34}$$

The combination of (2.33) and (2.34) allows the expression (2.20) for virtual work to take the form

$$\delta = \int_V t_{ij}\delta\varepsilon_{ij}\, dV. \tag{2.35}$$

*Editor's note: The idea is not discussed any further.

Under the hypothesis of geometric linearity, the linearized strain is conjugate to any of the stress measures, all of which become equivalent and symmetric.

# Chapter 3
# Conservation of Energy

## 3.1. Constitutive Equations for Piola's Stresses

In material coordinates, as we have seen, the density of work done by external forces in an infinitesimal change of configuration may be expressed by

$$\rho_o g_j \delta u_j + D_i(t_{ij}\delta u_j).$$

We assume that these changes occur at negligible velocities so as to ignore kinetic energy, and at uniform and constant temperature, with the necessary heat flux having sufficient time to occur. We thus require the existence of a local thermodynamic equilibrium, and equate the density of work by exterior forces to the change in an energy density $W$:

$$\delta W = \rho_o g_j \delta u_j + D_i(t_{ij}\delta u_j). \tag{3.1}$$

$W$ is called the *strain energy density*; from the thermodynamical point of view it is a free energy.

If on the other hand the changes occur so rapidly that the heat exchange between particles has no time to occur, the local thermodynamic equilibrium calls into play the density $U$ of internal energy, and the energy balance should also include changes in kinetic energy:

$$\rho_0 v_j \delta v_j = -\delta U + \rho_0 g_j \delta u_j + D_i(t_{ij}\delta u_j),$$
$$v_j = D_t u_j.$$

This equation agrees with the principle of rational mechanics according to which the kinetic energy of a collection of particles is the sum of the work done by external as well as internal forces $(-\delta U)$. It forms the foundation of elastodynamics.

Finally, the consideration of thermal exchanges during a global disturbance is the object of thermoelasticity.

We develop here only the consequences of the equation (3.1) of elastostatics. It contains virtually all the earlier developments concerning Piola's stresses, for its right hand side is nothing but the density of virtual work, from which analytical statics has enabled us to recover the translational and rotational equilibrium equations (2.8) and (2.10). In fact, the theorem of virtual work expressed by (2.20) and (2.21) can now be transformed into a theorem of conservation of energy by the equality

$$\delta = \int_V \delta W \; dV, \tag{3.2}$$

whence

$$\delta W = t_{ij}\delta a_{ji}. \tag{3.3}$$

This equation should be interpreted as the development of an exact differential for the strain energy density $W$. The function $W$ depends on the nine displacement gradients $a_{ij}$, and Piola's stresses form its partial derivatives:

$$t_{ij} = \frac{\partial W}{\partial a_{ji}} . \tag{3.4}$$

In (3.4) we have the most general form of the constitutive equations of the theory of elasticity.

Equation (3.4) is not perfect because it relates stresses to displacement gradients, the latter involving rotations as well as deformations. One may notice that it has been obtained without considering the rotational equilibrium equations. These introduce constraints responsible for a certain structure of $W$. Indeed, one can put equations (2.9) for rotational equilibrium into the form

$$e_{mnj}(t_{nj} + t_{ij}a_{ni}) = 0,$$

or

$$e_{mnj}\left(\frac{\partial W}{\partial a_{jn}} + a_{ni}\frac{\partial W}{\partial a_{ji}}\right) = 0 .$$

These are three partial differential equations which the function $W$ must satisfy.

For $m = 1$ the characteristics of the partial differential equation

$$\frac{\partial W}{\partial a_{32}} + a_{2i}\frac{\partial W}{\partial a_{3i}} - \frac{\partial W}{\partial a_{23}} - a_{3i}\frac{\partial W}{\partial a_{2i}} = 0$$

have the equations

$$\frac{da_{12}}{0} = \frac{da_{23}}{-(1+a_{33})} = \frac{da_{31}}{a_{21}} = \frac{da_{21}}{-a_{31}} = \frac{da_{32}}{1+a_{22}} = \frac{da_{13}}{0}$$

$$= \frac{da_{11}}{0} = \frac{da_{22}}{-a_{32}} = \frac{da_{33}}{a_{23}} = \frac{dW}{0} .$$

These show that $W$ should be a function of the first integrals of the differential equations connecting the $a_{ij}$. Immediate first integrals are

$$a_{11}\ , \quad a_{12}\ , \quad a_{13}\ ,$$
$$a_{21}^2 + a_{31}^2, \quad a_{22} + \tfrac{1}{2}(a_{22}^2 + a_{32}^2), \quad a_{33} + \tfrac{1}{2}(a_{33}^2 + a_{23}^2),$$

and

$$a_{22}(1 + a_{22}) + a_{32}(1 + a_{33}),$$
$$a_{31}(1 + a_{33}) + a_{23}a_{21},$$
$$a_{21}(1 + a_{22}) + a_{32}a_{31}.$$

In other words, these quantities are arbitrary constants for the differential connections of the characteristics. One verifies that the six quantities

$$\varepsilon_{ij} = \tfrac{1}{2}(a_{ij} + a_{ji} + a_{mi}a_{mj}) = \varepsilon_{ji} \tag{3.5}$$

are precise combinations of these first integrals and hence, because of their symmetry, they are also first integrals of the characteristics of the other two partial differential equations for $m = 2$ and $m = 3$.

Indeed, the expressions (3.5) are the only independent first integrals common to the three systems. Because of the definitions (2.21) and (1.19), the $\varepsilon_{ij}$ are merely the components of Green's strain, and we obtain the result that the Lagrangian strain energy density $W$ is a function of the nine components of displacement gradient only through the six distinct elements forming the symmetric matrix of Green's strain.

## 3.2. The Kirchhoff-Trefftz Stresses

The foregoing result allows the formulation of a natural energetic definition of new stresses, of a purely Lagrangian nature, considered by *Piola* and *Kirchhoff*, of which *Trefftz* has shown the importance and utility.

For generality of indicial formulation, the function $W$ of the six distinct $\varepsilon_{ij}$ will be considered as a function of the six arguments $\frac{1}{2}(\varepsilon_{ij} + \varepsilon_{ji})$, where one maintains a formal distinction between $\varepsilon_{ij}$ and $\varepsilon_{ji}$. Thus we obtain formally nine partial derivatives

$$s_{ij} = \frac{\partial W}{\partial \varepsilon_{ij}} \tag{3.6}$$

but the symmetrization of the arguments of the function has the consequences

$$s_{ji} = \frac{\partial W}{\partial \varepsilon_{ji}} = \frac{\partial W}{\partial \varepsilon_{ij}} = s_{ij}. \tag{3.7}$$

Equations (3.6) are thus considered as general constitutive equations relating Green's strain to the *Kirchhoff-Trefftz stresses* $s_{ij}$ which are themselves symmetric. The relation between these new stresses and Piola's follow directly from the chain rule:

$$t_{qp} = \frac{\partial W}{\partial a_{pq}} = \frac{\partial W}{\partial \varepsilon_{ij}} \frac{\partial \varepsilon_{ij}}{\partial a_{pq}} = s_{ij} \frac{\partial \varepsilon_{ij}}{\partial a_{pq}} .$$

From (3.5) we obtain

$$\frac{\partial \varepsilon_{ij}}{\partial a_{pq}} = \frac{1}{2}(\delta_{ip}\delta_{jq} + \delta_{jp}\delta_{iq} + \delta_{mp}\delta_{iq}a_{mj} + \delta_{jq}\delta_{mp}a_{mi}),$$

and consequently,

$$t_{pq} = \frac{1}{2}(s_{pq} + s_{qp} + s_{qj}a_{pj} + s_{iq}a_{pi}).$$

By using the symmetry of $s_{pq}$, we finally obtain

$$t_{pq} = s_{qp} + s_{qj}a_{pj} = s_{qj}(\delta_{pj} + D_j u_p) = s_{qj}D_j y_p. \tag{3.8}$$

It is easy to verify that this relation does satisfy the conditions (2.10) for rotational equilibrium of Piola's stresses.

By substituting (3.8) into the equivalent forms

$$t_{qp}D_q y_n = t_{qn}D_q y_p,$$

we obtain

$$s_{qj}D_q y_n D_j y_p = s_{qj}D_j y_n D_q y_p.$$

Exchanging the indices $q$ and $j$ in the right hand side yields

$$s_{qj}D_q y_n D_j y_p = s_{jq}D_q y_n D_j y_p.$$

These equations are valid because of the symmetry of the Kirchhoff-Trefftz stresses. We may thus consider equations (3.8) as furnishing a definition of these stresses:

$$s_{qr} = \partial_p x_r t_{qp} , \tag{3.9}$$

enabling the replacement of the conditions of rotational equilibrium by a simple symmetry. The formula (3.3) for conservation of energy thus takes the form

$$\delta W = s_{im}D_m y_j \delta D_i u_j = s_{im}D_m y_j \delta D_i y_j .$$

By interchanging the indices $i$ and $m$ and using the symmetry $s_{mi} = s_{im}$, it becomes

$$\delta W = s_{im}D_i y_j \delta D_m y_j ,$$

or

$$\begin{aligned}\delta W &= s_{im}\, \tfrac{1}{2}(D_m y_j \delta D_i y_j + D_i y_j \delta D_m y_j) \\ &= s_{im}\, \tfrac{1}{2}\delta(D_m y_j D_i y_j) = s_{im}\, \tfrac{1}{2}\delta(2\varepsilon_{im} - \delta_{im}),\end{aligned}$$

and finally

$$\delta W = s_{im}\delta\varepsilon_{im}. \tag{3.10}$$

This formula is equivalent to the preceding definition (3.6).

Trefftz has given an elegant geometric interpretation of the relations (3.8). We recall that the contact force on a surface element is expressed as a function of Piola's stresses by

$$d\mathbf{F} = dS\, n_q t_{qp} \mathbf{e}_p.$$

Substitution of (3.8) yields

$$d\mathbf{F} = dS\, n_q s_{qj} D_j y_p \mathbf{e}_p.$$

We have seen that

$$D_j y_p \mathbf{e}_p = \mathbf{g}_j$$

are the basis vectors of the system of natural curvilinear coordinates generated by the convection of the Cartesian coordinate planes of the initial configuration. By rewriting $d\mathbf{F}$ as

$$d\mathbf{F} = dS\, n_q s_{qj} \mathbf{g}_j, \tag{3.11}$$

one observes that the Kirchhoff-Trefftz stresses are forces per unit of initial surface, but result from a decomposition of the surface traction in the natural basis induced by the change of configuration.

## 3.3. The Constitutive Equations of Geometrically Linear Elasticity

Equation (2.35) for the virtual work in an infinitesimal change of configuration, when compared with equation (3.2) for conservation of energy, directly yields the result

$$\delta W = t_{ij} \delta \varepsilon_{ij}. \tag{3.12}$$

In this formula, $\varepsilon_{ij}$ denotes the infinitesimal strain given by (2.34), and $t_{ij}$, any one of the stress measures, which, under the hypothesis of geometric linearity, are all symmetric and practically equivalent with each other. The detailed form of the stress-strain law arising from (3.12) will be discussed at length in Chapter 5.

# Chapter 4
# Cartesian Tensors

## 4.1. Bases and Change of Basis

Let $\mathbf{e}_i$ be the unit vectors of the Cartesian frame. They constitute an orthonormal basis, i.e.,

$$\mathbf{e}_i \cdot \mathbf{e}_j = \delta_{ij}, \tag{4.1}$$

and in this basis each vector $\mathbf{u}$ has the unique representation

$$\mathbf{u} = u_i \mathbf{e}_i \tag{4.2}$$

in terms of its components $u_i$. The orthogonal projections of the vector on the axes,

$$\mathbf{u} \cdot \mathbf{e}_j = u_i \mathbf{e}_i \cdot \mathbf{e}_j = u_i \delta_{ij} = u_j, \tag{4.3}$$

are identical with its components. As long as we use only Cartesian frames we do not need the distinction between contravariant components, usually defined by (4.2), and covariant components, usually defined by (4.3).

Let $\mathbf{e}'_k$ be another orthonormal basis. It is fully determined with respect to the original frame by the nine numbers

$$T_{ki} = \mathbf{e}'_k \cdot \mathbf{e}_i. \tag{4.4}$$

Each $T_{ki}$ is the cosine of the angle formed by the vectors $\mathbf{e}'_k$ and $\mathbf{e}_i$. If the $T_{ki}$ are arranged in a matrix $T$ so that $k$ is the row index and $i$ the column index, a row of $T$ gives the components of $\mathbf{e}'_k$ with respect to the $\mathbf{e}_i$, while a column of $T$ gives the components of $\mathbf{e}_i$ with respect to the $\mathbf{e}'_k$.

The nine $T_{ki}$ are clearly not independent. Since

$$\mathbf{e}'_k = T_{ki}\mathbf{e}_i, \tag{4.5}$$

the fact that the new frame is also orthonormal implies

$$\mathbf{e}'_k \cdot \mathbf{e}'_\ell = T_{ki}T_{\ell j}\mathbf{e}_i \cdot \mathbf{e}_j = T_{ki}T_{\ell i} = \delta_{k\ell}. \tag{4.6}$$

In matrix notation, (4.6) becomes

$$TT^T = I. \tag{4.7}$$

By taking determinants on each side, we obtain

$$\det T \cdot \det T^T = (\det T)^2 = 1. \tag{4.8}$$

Multiplication of (4.7) on the right by $T$ yields, after regrouping,

$$T(T^TT - I) = 0.$$

The matrix $T$ is nonsingular because its determinant is $\pm 1$, so we obtain

$$T^TT = I \quad \text{or} \quad T_{ki}T_{kj} = \delta_{ij}. \tag{4.9}$$

Thus multiplication of (4.5) by $T_{kj}$ yields the formula for the inverse change of basis

$$\mathbf{e}_j = T_{kj}\mathbf{e}'_k. \tag{4.10}$$

This relation also follows from the interpretation already

given for the columns of $T$.

## 4.2. Tensors

We may refer Euclidean space to any one of an infinity of frames (coordinate systems) which we regard as fixed with respect to each other. A physical quantity will be called a *tensor* if one may verify its independence of the frame in use. This independence is manifested by the invariance of a number or a set of numbers under a change of basis. We shall consider only Cartesian frames; the invariance then characterizes *Cartesian* tensors.

A physical quantity such as the density $\rho$ at a point and for an arbitrarily chosen configuration of the continuous medium is expressed by a single number, which does not depend on the frame in use. Such a scalar invariant is thus, by definition, a tensor of order zero.

A fixed direction in Euclidean space indicated by a vector $\mathbf{n}$ (preferably but not necessarily of unit length) is a tensor of order one.* In a Cartesian frame with basis $\mathbf{e}_i$ this direction is represented by three numbers $n_i$, the Cartesian components of this vector:

$$\mathbf{n} = n_i \mathbf{e}_i. \tag{4.11}$$

In another frame $\mathbf{e}'_k$ the same direction will be represented by three other numbers $n'_k$:

$$\mathbf{n} = n'_k \mathbf{e}'_k. \tag{4.12}$$

---

*Editors note: The manipulations that follow are valid for any physical quantity characterized by a magnitude and a direction, and are not restricted to unit vectors.

By combining (4.5), (4.11), and (4.12), we obtain

$$n = n'_k e'_k = n'_k T_{ki} e_i = n_i e_i.$$

Thus under a change of coordinates, the components of a tensor of order one, or a *polar vector*, transform according to

$$n_i = n'_k T_{ki}. \tag{4.13}$$

Multiplication of (4.13) on the right by $T_{\ell i}$ yields the inverse relation

$$n_i T_{\ell i} = n'_k T_{ki} T_{\ell i} = n'_k \delta_{k\ell} = n'_\ell. \tag{4.14}$$

One method for recognizing the tensorial character of a quantity represented is to verify if its components transform according to (4.13) or (4.14) under a change of basis.

Another method, with a simple generalization to tensors of higher order, consists in verifying the invariance of the scalars formed with one or more arbitrary but fixed directions in Euclidean space. Let $u_i$ be the numbers representing in a basis $e_i$ a quantity whose character as a tensor of order one is to be verified. Consider the linear form

$$\phi = u_i n_i$$

where $n_i$ represent a fixed but arbitrary direction in space. In a new basis we have

$$\phi' = u'_k n'_k = u'_k T_{ki} n_i.$$

If the value of the form is invariant, then $\phi' = \phi$ and because of the arbitrary character of the chosen orientation, we obtain by comparison

$$u_i = u'_k T_{ki}.$$

The invariance of $\phi$ thus implies that the transformation law of type (4.13) holds and consequently that the numbers $u_i$ are the components of a tensor of order one. Conversely, if the $u_i$ are components of a tensor of order one, then the linear form is invariant. The interpretation of the invariant linear form is clear:

$$\phi = u_i n_i = \mathbf{u} \cdot \mathbf{n}$$

is the scalar product of two polar vectors; it depends only on the lengths of the vectors and the angle between them.

A Cartesian tensor of order two is associated with the invariance of the bilinear form

$$\phi = t_{ij} n_i m_j$$

containing two arbitrary directions. Invariance requires that

$$t_{ij} n_i m_j = t_{ij} T_{ki} T_{\ell j} n'_k m'_\ell = t'_{k\ell} n'_k m'_\ell$$

for every choice of direction; we thus have the transformation

$$t'_{k\ell} = T_{ki} T_{\ell j} t_{ij} \quad \text{and} \quad t_{ij} = T_{ki} T_{\ell j} t'_{k\ell}. \tag{4.15}$$

The generalization to tensors of higher order should be clear.

Tensors of order zero require no special symbol. Tensors of order one, or polar vectors, enjoy a special symbol such as $\mathbf{u}$. As a notation for tensors of higher order, we shall use indicial notation for the components, and not yet adopt a special symbol for the tensor as a whole, although one would prefer to avoid indicial notation with its intrinsic reference to a particular frame. A natural abuse of language is allowed in speaking of the tensor $t_{ij}$, although the numbers $t_{ij}$ constitute in fact the representation of the tensor only with respect to the particular frame in use. The situation is the same if one defines a tensor as an invariant multilinear form of several directions in space. The form is characterized by its coefficients, which are the components of the tensor in a particular frame.

## 4.3. Some Special Tensors

The scalar product of two polar vectors can be expressed as a bilinear form in the two directions of its factors:

$$\mathbf{m} \cdot \mathbf{n} = m_i n_i = \delta_{ij} n_i n_j .$$

The coefficients of this form are Kronecker's symbols $\delta_{ij}$. The form being invariant, the Kronecker symbol is a Cartesian tensor of order two. It has the remarkable property of being *isotropic*, or *spherical*. This means that its components have the same value in any frame. Indeed, the components $\delta'_{k\ell}$ in a new frame are, according to (4.15),

$$\delta'_{k\ell} = T_{ki} T_{\ell j} \delta_{ij} = T_{ki} T_{\ell i} = \delta_{k\ell}$$

which, in view of (4.6), gives $\delta_{k\ell}$ its meaning as Kronecker's symbol in the new indices, and renders the superscript unnecessary. One may show that except for modulus, $\delta_{ij}$ is the only isotropic Cartesian tensor of the second order.

We come now to the permutation symbol $e_{ijm}$, defined on page 16. Consider the trilinear form in three directions, whose coefficients are the permutation symbols:

$$\phi = e_{ijm} m_i n_j p_m . \tag{4.16}$$

It follows from the Laplace expansion of a determinant that

$$\phi = \det\begin{pmatrix} m_1 & m_2 & m_3 \\ n_1 & n_2 & n_3 \\ p_1 & p_2 & p_3 \end{pmatrix}.$$

When the directions are expressed in a new frame one finds

$$\phi = e_{ijm}T_{ki}T_{\ell j}T_{nm}m'_k n'_\ell p'_n,$$

but, again by Laplace's formula,

$$e_{ijm}T_{ki}T_{\ell j}T_{nm} = e_{k\ell n}\det T \tag{4.17}$$

where $e_{k\ell n}$ are as defined on page 16. In view of (4.8), we must distinguish two cases:

First, suppose $\det T = +1$. Then, because the left hand side of (4.17) is the formula for the transformed components $e'_{k\ell n}$ of $e_{ijm}$ considered as a tensor of order three, with $e'_{k\ell n} = e_{k\ell n}$ the trilinear form is invariant:

$$\phi = e_{k\ell n}m'_k n'_\ell p'_n = \phi'.$$

When $\det T = +1$, the change of the Cartesian frame conserves the sense of the frame (right-handed or left-handed), or in other words, the new frame can be obtained from the old by a rotation. If only such transformations are admitted, the permutation symbol may be regarded as an isotropic Cartesian tensor of order three. It is antisymmetric under the interchange of any two indices.

Second, suppose $\det T = -1$. We then have a change of frame which alters its sense: In order to obtain the new frame from the old, we must perform a rotation and a reflection in a plane. If one does not wish to ignore this case, then formula (4.17) for change of components may be regarded as defining a *pseudo-tensor*. One may then say that the permutation symbol is an isotropic Cartesian pseudo-tensor of order three. The associated trilinear form takes a sign connected with the sense of the frame in use.

Although in principle the permutation symbol has $3^3 = 27$ components, it is really defined, because of its antisymmetry, by the choice of a single value. This is conventionally chosen as $e_{123} = 1$ and is called the *strict component*. Because the permutation symbol, considered as a pseudo-tensor, is isotropic, its strict component is invariant and takes the numerical value of the associated tensor of order zero.

It is clear that one may equally well define a genuinely antisymmetric tensor of order three, which we denote by $\varepsilon_{ijk}$ to distinguish it from the permutation symbol. Its strict component $\varepsilon_{123}$ is set equal to $+1$ for a frame of conventionally agreed orientation, and it is the associated tensor of order zero which becomes a *pseudo-scalar*, its sign changing with the orientation of the frame.

Among the isotropic tensors of order four, the most important is that obtained by contraction

$$e_{ijk}e_{pqk} = \delta_{ip}\delta_{jq} - \delta_{iq}\delta_{jp} = \begin{Bmatrix} i & j \\ p & q \end{Bmatrix} . \tag{4.18}$$

In order to prove this extremely useful formula, one may use the formula

$$\det\begin{pmatrix} \delta_{1i} & \delta_{2i} & \delta_{3i} \\ \delta_{1j} & \delta_{2j} & \delta_{3j} \\ \delta_{1k} & \delta_{2k} & \delta_{3k} \end{pmatrix} = e_{ijk}.$$

Then, because the determinant of a matrix equals that of its transpose, and the product of the determinants of two matrices equals the determinant of their product, one has

$$e_{ijk}e_{pqk} = \det\left\{\begin{pmatrix}\delta_{1i} & \delta_{2i} & \delta_{3i}\\ \delta_{1j} & \delta_{2j} & \delta_{3j}\\ \delta_{1k} & \delta_{2k} & \delta_{3k}\end{pmatrix}\begin{pmatrix}\delta_{1p} & \delta_{1q} & \delta_{1k}\\ \delta_{2p} & \delta_{2q} & \delta_{2k}\\ \delta_{3p} & \delta_{3q} & \delta_{3k}\end{pmatrix}\right\}$$

$$= \det\begin{pmatrix}\delta_{ip} & \delta_{iq} & \delta_{ik}\\ \delta_{jp} & \delta_{jq} & \delta_{jk}\\ \delta_{kp} & \delta_{kq} & \delta_{kk}\end{pmatrix} .$$

By noting that $\delta_{kk} = 3$ and developing the determinant by elements of the last column, we arrive at (4.18). For $q = j$ it reduces to

$$e_{ijk}e_{pjk} = 2\delta_{ip}, \tag{4.19}$$

and for $q = j$ and $i = p$ one finds

$$e_{ijk}e_{ijk} = 2\delta_{ii} = 6. \tag{4.20}$$

## 4.4. The Vector Product

The *vector product* of two vectors $\mathbf{u}$ and $\mathbf{v}$ is written $\mathbf{u} \times \mathbf{v}$, and its general definition follows from the restricted definition for the basis vectors:

$$\mathbf{e}_i \times \mathbf{e}_j = e_{ijk}\mathbf{e}_k. \tag{4.21}$$

From the hypotheses of associativity and distributivity, there then follows

$$\mathbf{u} \times \mathbf{v} = u_i v_j \mathbf{e}_i \times \mathbf{e}_j = e_{ijk}u_i v_j \mathbf{e}_k. \tag{4.22}$$

In detail, the components are

$$(u \times v)_1 = u_2v_3 - u_3v_2,$$

$$(u \times v)_2 = u_3v_1 - u_1v_3,$$

$$(u \times v)_3 = u_1v_2 - u_2v_1.$$

Because of the permutation symbol in the definition, the vector product of two polar vectors is a *pseudo-vector*. Indeed, by the definition (4.21) in the new frame,

$$\mathbf{e}'_k \times \mathbf{e}'_\ell = e_{k\ell n}\mathbf{e}'_n$$

and, by (4.17) and (4.10),

$$\mathbf{e}'_k \times \mathbf{e}'_\ell = \frac{1}{\det T} T_{ki}T_{\ell j}T_{nm}e_{ijm}\mathbf{e}'_n = \frac{1}{\det T} T_{ki}T_{\ell j}e_{ijm}\mathbf{e}_m,$$

or, by virtue of (4.21) in the old frame,

$$T_{ki}T_{\ell j}\mathbf{e}_i \times \mathbf{e}_j = \det T\, \mathbf{e}_k \times \mathbf{e}_\ell\ . \qquad (4.23)$$

This law of pseudo-tensorial transformation for the vector products of the basis evidently implies the same law for the vector product of any two polar vectors.

## 4.5. Structure of Symmetric Cartesian Tensors of Order Two. Principal Axes

The symmetry of a Cartesian tensor with respect to two of its indices is an intrinsic property. Indeed, if

$$t_{ij} = t_{ji} \qquad (4.24)$$

in a given frame, then

$$t'_{k\ell} = T_{ki}T_{\ell j}t_{ij} = T_{kj}T_{\ell i}t_{ji} = T_{kj}T_{\ell i}t_{ij} = t'_{\ell k}.$$

The first equality results from the law of tensorial transformation, the second involves only the interchange of the indices of summation (dummy indices), the third applies the symmetry assumed in the original frame, and the last uses again the law of tensorial transformation.

A symmetric tensor being a physical quantity independent of the frame, one may ask if it is possible to associate favored directions in space with it. The tensor may be regarded as an operator (viz., a linear transformation) establishing a correspondence between vectors. Let

$$t_{ij}u_j = v_i$$

be the correspondence $\mathbf{u} \to \mathbf{v}$ referred to the frame $\mathbf{e}_i$. We now inquire as to the existence of a favored direction in space, invariant under the correspondence, and a scalar $\tau$, called an *eigenvalue* of the tensor, such that if $\mathbf{u}$ lies along that direction,

$$\mathbf{v} = \tau\, \mathbf{u}.$$

(Since we allow negative as well as positive real values for $\tau$, we agree to speak of invariance of *direction*, while invariance of *orientation* would require that $\tau > 0$).

The problem leads to seeking solutions of the algebraic problem

$$t_{ij}u_j = \tau\, u_i \quad \text{or} \quad (t_{ij} - \tau\delta_{ij})u_j = 0. \tag{4.25}$$

This linear homogeneous system in the unknowns $u_i$ has a nontrivial solution only if

$$\det(t_{ij} - \tau\delta_{ij}) = 0. \tag{4.26}$$

This is the equation for the eigenvalues of the tensor. They are independent of the frame because, in another frame,

$$\det(t'_{k\ell} - \tau\delta'_{k\ell}) = \det(T_{ki}T_{\ell j}t_{ij} - \tau T_{ki}T_{\ell j}\delta_{ij})$$
$$= \det T_{ki} \det T_{\ell j} \det(t_{ij}-\tau\delta_{ij}) = \det(t_{ij} - \tau\delta_{ij}).$$

The first equality uses the tensorial character of $t_{ij}$ and $\delta_{ij}$, the second uses the theorem on the determinant of a product of matrices, and the third applies (4.8).

It is not impossible, a priori, that an eigenvalue, a root of (4.26), might be a complex number. If this is so, then equation (4.25) may require that the solution $u_i$, which is defined only up to a real or complex factor, consist also of complex numbers. If $u_i^*$ are the conjugate complex numbers, we have the equality

$$u_i^* t_{ij} u_j = \tau u_i^* u_i . \tag{4.27}$$

Being identical with its conjugate, $u_i^* u_i$ is a real number; the same is true of $u_i^* t_{ij} u_j$, for its conjugate

$$u_i t_{ij}^* u_j^* = u_j t_{ji}^* u_i^* = u_j t_{ij} u_i^*$$

may be reduced to the initial expression by first interchanging the dummy indices, then observing that the symmetry (4.24) in real numbers is equivalent to $t_{ji}^* = t_{ij}$. It then follows from (4.27) that $\tau$, a quotient of two real numbers whose denominator vanishes only if $u_1 = u_2 = u_3 = 0$, is itself a real number. Thus the symmetry of the tensor has the important consequence that its eigenvalues must be real. To each real eigenvalue there corresponds, accordingly, at least one nontrivial real solution of the algebraic system (4.25). Being defined only up to a real factor, it identifies only a favored direction in space, and not an orientation.

Being of the third degree, equation (4.26) for the eigenvalues admits, in principle, three roots. We now show that if two roots are distinct, the corresponding orientations are orthogonal. For the first root $\tau_1$, let $u_i^{(1)}$ be an associated proper solution, and $u_i^{(2)}$ for $\tau_2$. By definition,

$$t_{ij}u_j^{(1)} = \tau_1 u_i^{(1)}$$

and, therefore,

$$t_{ij}u_i^{(2)}u_j^{(1)} = \tau_1 u_i^{(1)}u_i^{(2)}. \tag{4.28}$$

Similarly,

$$t_{ij}u_j^{(2)} = \tau_2 u_i^{(2)}$$

and, consequently,

$$t_{ij}u_j^{(2)}u_i^{(1)} = t_{ji}u_i^{(2)}u_j^{(1)} = \tau_2 u_i^{(1)}u_i^{(2)}.$$

By subtracting this last relation member by member from the equality (4.28) and observing that the symmetry (4.24) will annul the first term of the result, we have

$$0 = [\tau_1 - \tau_2]u_i^{(1)}u_i^{(2)}.$$

Since by hypothesis the eigenvalues are distinct, we obtain the orthogonality relation

$$u_i^{(1)}u_i^{(2)} = 0 \quad \text{if} \quad \tau_1 \neq \tau_2. \tag{4.29}$$

If the three eigenvalues are distinct, there exists a set of three mutually orthogonal favored directions, called the principal axes or *eigenvectors* of the tensor. As they are defined only up to a real factor, we may make them unit vectors by normalizing them such that

$$u_i^{(m)}u_i^{(m)} = 1, \qquad m = 1,2,3, \text{ no sum on } m.$$

We use these three unit vectors as a new orthonormal frame. Referred to it, the components of the eigenvectors associated with $\tau_1$, $\tau_2$, and $\tau_3$ are respectively (1,0,0), (0,1,0), and (0,0,1). In this frame the algebraic system (4.25) has as one solution

$$t_{11} = \tau_1 \qquad t_{21} = 0 \qquad t_{31} = 0. \tag{4.30}$$

For the second proper solution we choose

$$t_{12} = 0 \qquad t_{22} = \tau_2 \qquad t_{32} = 0$$

and for the third

$$t_{13} = 0 \qquad t_{23} = 0 \qquad t_{33} = \tau_3.$$

It is clear that the matrix representing the tensor in a frame of principal axes is diagonal and its diagonal elements are the eigenvalues:

$$\begin{pmatrix} \tau_1 & 0 & 0 \\ 0 & \tau_2 & 0 \\ 0 & 0 & \tau_3 \end{pmatrix}. \tag{4.31}$$

We are left with degenerate cases, those corresponding to a multiple root of the equation for eigenvalues. In the case of a double root, let $\tau_1$ be the simple root. Adopt a frame whose first vector is one of the normed eigenvectors $u^{(1)}$ (there are clearly two possibilities corresponding to the two directions along the first principal axis). In such a frame (4.30) holds and, since symmetry is an intrinsic property, the matrix representing the tensor in the new frame will have the form

$$\begin{pmatrix} \tau_1 & 0 & 0 \\ 0 & t_{22} & t_{23} \\ 0 & t_{23} & t_{33} \end{pmatrix}$$

with $t_{32} = t_{23}$. After developing the equation for the eigenvalues in the new frame, the known solution $\tau = \tau_1$ may be factored out and the two others become roots of

$$\det\begin{vmatrix} t_{22}-\tau & t_{23} \\ t_{23} & t_{33}-\tau \end{vmatrix} = \tau^2 - \tau(t_{22}+t_{33}) + t_{22}t_{33} - t_{23}^2 = 0$$

(Note that the discriminant of this algebraic equation is

$$(t_{22}+t_{33})^2 - 4(t_{22}t_{33}-t_{23}^2) = (t_{22}-t_{33})^2 + 4t_{23}^2 \geq 0,$$

confirming that the other two roots are real.) The condition for a double root, the vanishing of the discriminant, requires simultaneously

$$t_{22} = t_{33} \ (= \tau_2) \quad \text{and} \quad t_{23} = t_{32} = 0.$$

Then the tensor is easily seen to have a diagonal structure

$$\begin{pmatrix} \tau_1 & 0 & 0 \\ 0 & \tau_2 & 0 \\ 0 & 0 & \tau_2 \end{pmatrix}$$

and all the axes perpendicular to $u^{(1)}$ are principal. Such a tensor is said to be *cylindrical*. For a double root, therefore, the algebraic system (4.25) has two linearly independent solutions, and the general solution is an arbitrary linear combination of them.

Finally, if the equation for the eigenvalues has a triple root, one may always determine a solution of (4.25) and then a principal axis. By virtue of the preceding, we then have the structure

$$\tau\begin{pmatrix}1 & 0 & 0\\ 0 & 1 & 0\\ 0 & 0 & 1\end{pmatrix} \quad \text{or} \quad t_{ij} = \tau\delta_{ij} \tag{4.32}$$

in every frame with the first vector pointing in that direction. We have seen that $\delta_{ij}$ is an isotropic tensor and, therefore, differing only by the modulus $\tau$, $t_{ij}$ is itself an isotropic or spherical tensor, with the same structure in all Cartesian frames. For such a tensor, all directions in space are principal.

## 4.6. Fundamental Invariants and the Deviator

The eigenvalues being independent of the frame, so are the coefficients of the algebraic equation (4.6), which are symmetric functions of the roots. These coefficients are the *fundamental invariants* of the tensor. Expanding (4.26) yields

$$\phi(\tau) = -\tau^3 + \theta_1\tau^2 - \theta_2\tau + \theta_3 = (\tau_1-\tau)(\tau_2-\tau)(\tau_3-\tau) = 0. \tag{4.33}$$

The fundamental invariants $\theta_i$ are expressed in terms of the eigenvalues $\tau_i$ as follows:

$$\begin{aligned}\theta_1 &= \tau_1 + \tau_2 + \tau_3\\ \theta_2 &= \tau_1\tau_2 + \tau_2\tau_3 + \tau_2\tau_1\\ \theta_3 &= \tau_1\tau_2\tau_3.\end{aligned} \tag{4.34}$$

They may also be written as functions of the components of the tensor in an arbitrary frame. For this purpose we may return again to the Laplace's formula; in view of (4.20) one may write

$$e_{ijk}(t_{im}-\tau\delta_{im})(t_{jn}-\tau\delta_{jn})(t_{kp}-\tau\delta_{kp}) = e_{mnp}\det(t_{ij} - \tau\delta_{ij}).$$

Expanding and reducing yields

$$\theta_1 = \frac{1}{2} e_{mnp}e_{inp}t_{im} = \delta_{im}t_{im} = t_{ii} \tag{4.35}$$

$$\begin{aligned}\theta_2 &= \frac{1}{2} e_{mnp}e_{mjk}t_{nj}t_{pk} = \frac{1}{2}(t_{jj}t_{kk} - t_{kj}t_{jk}) \\ &= t_{22}t_{33}+t_{33}t_{11}+t_{11}t_{22}-(t_{23}^2+t_{31}^2+t_{12}^2)\end{aligned} \tag{4.36}$$

$$\theta_3 = \frac{1}{6} e_{mnp}e_{ijk}t_{im}t_{jn}t_{kp} = \det(t_{ij}). \tag{4.37}$$

Let us calculate the changes in the fundamental invariants when the tensor is modified by subtracting a spherical tensor. Thus we shall evaluate the fundamental invariants of the modified tensor

$$t_{ij} - \mu\delta_{ij}.$$

The equation for the eigenvalues is obtained by replacing $\tau$ by $\tau+\mu$ in (4.33) whence, after reordering according to powers of $\tau$,

$$\begin{aligned}-\tau^3 + \tau^2(\theta_1-3\mu) - \tau(\theta_2+3\mu^2-2\theta_1\mu) \\ + \theta_3 - \mu^3 + \theta_1\mu^2 - \theta_2\mu = 0.\end{aligned}$$

The new second fundamental invariant

$$\hat{\theta}_2 = \theta_2 + 3\mu^2 - 2\theta_1\mu \equiv -\phi'(\mu)$$

is maximized when $\mu = \theta_1/3$, and this value of $\mu$ annuls the new first fundamental invariant. Moreover, for this maximum we have

$$\hat{\theta}_2 = \theta_2 - \frac{1}{3}\theta_1^2 = -\phi'(\frac{1}{3}\theta_1) \leq 0, \tag{4.38}$$

as one sees at once from the graphs of $\phi(\tau)$ and of $\phi'(\tau)$ whose roots are known to be real. The maximum vanishes only if the three eigenvalues are equal, when the tensor under study

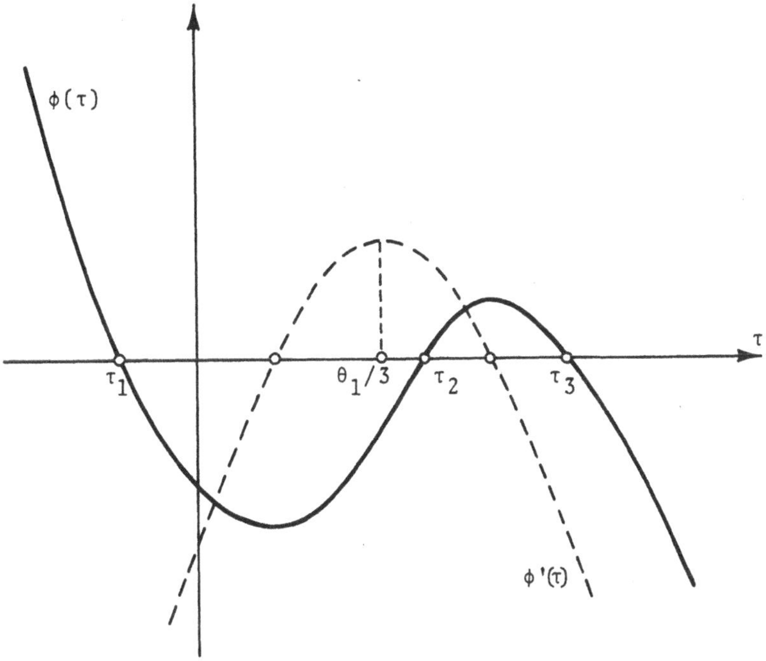


Figure 4.1

is already spherical.

This result is confirmed by expressing (4.38) in the components $t_{ij}$. From (4.35) and (4.36) we get

$$\hat{\theta}_2 = \frac{1}{6}(t_{jj}t_{kk} - 3t_{kj}t_{jk}), \tag{4.39}$$

or, after expansion,

$$-\hat{\theta}_2 = \frac{1}{6}[(t_{11}-t_{22})^2+(t_{22}-t_{33})^2+(t_{33}-t_{11})^2] + t_{12}^2 + t_{23}^2 + t_{31}^2 \geq 0. \tag{4.40}$$

One sees easily that if $\hat{\theta}_2$ vanishes then

$$t_{12} = t_{23} = t_{31} = 0 \qquad t_{11} = t_{22} = t_{33} = \frac{1}{3}\,\theta_1.$$

The tensor $\frac{1}{3}\,\theta_1\delta_{ij}$ is, by definition, the spherical part of the tensor $t_{ij}$, and its complement

$$\hat{t}_{ij} = t_{ij} - \tfrac{1}{3}\theta_1\delta_{ij}, \quad \hat{\theta}_1 = \hat{t}_{kk} = 0, \quad \hat{\theta}_2 = -\tfrac{1}{2}\hat{t}_{kj}\hat{t}_{jk} \tag{4.41}$$

is the *deviatoric* part, or the *deviator*, of the tensor $t_{ij}$.

The first fundamental invariant of the deviator vanishes. The second fundamental invariant is negative and its vanishing implies the complete disappearance of the deviator. One sees also that the vanishing of the first two invariants $(\theta_1, \theta_2)$ is a necessary and sufficient condition that the tensor $t_{ij}$ vanish identically. This is confirmed by the following relation, easily established using principal axes:

$$(\tau_1^2 + \tau_2^2 + \tau_3^2) = \theta_1^2 - 2\theta_2. \tag{4.42}$$

## 4.7. Structure of Skew-Symmetric Cartesian Tensors of the Second Order

For a skew-symmetric tensor, the symmetry relation (4.24) is replaced by that of skew-symmetry

$$k_{ji} = -k_{ij}. \tag{4.43}$$

Such a tensor has only three distinct components and we may always associate with it a pseudo-vector

$$k_m = -\tfrac{1}{2}\, e_{mnp}k_{np} \tag{4.44}$$

with components

$$\begin{aligned} k_1 &= -k_{23} = k_{32}, \\ k_2 &= -k_{31} = k_{13}, \\ k_3 &= -k_{12} = k_{21}. \end{aligned} \tag{4.45}$$

Inversely

$$k_{ij} = -e_{ijm}k_m. \tag{4.46}$$

By using (4.17), it is easy to verify that if $k_{ij}$ obeys the transformation laws of a genuine tensor of the second order then $k_m$ obeys those of a pseudo-tensor

$$k_m = T_{\ell m} k'_\ell \det T. \tag{4.47}$$

Every skew tensor of the second order has an eigenvalue which vanishes. Indeed, the equation

$$(k_{ij} - \kappa\delta_{ij})u_j = 0$$

is satisfied by $u_j = k_j$ and $\kappa = 0$, because

$$k_{ij}k_j = -e_{ijm}k_j k_m = 0, \tag{4.48}$$

the terms cancelling two by two. We see that $\kappa = 0$ is the only real eigenvalue. Consequently, the tensor has only one principal direction, that of its associated pseudo-vector. For reasons connected with a possible kinematic interpretation of the tensor as an operator, we call this direction the *axis of rotation* of the tensor.

In a frame with first basis vector aligned with the axis of rotation, the matrix of the tensor takes the form

$$\begin{pmatrix} 0 & 0 & 0 \\ 0 & 0 & -k_1 \\ 0 & k_1 & 0 \end{pmatrix}$$

(This result could have been obtained from (4.48) and the fact that skew-symmetry is itself an intrinsic property.) The equation for eigenvalues, always invariant, reduces to

$$\kappa(\kappa^2 + k_1^2) = 0.$$

Along with the zero root, the tensor has a pair of conjugate complex eigenvalues $\kappa = \pm ik_1$, $(i = \sqrt{-1})$.

Except for a factor, the corresponding eigenvectors may be chosen as

$$w_1 = 0, \quad w_2 = 1, \quad w_3 = -i, \quad \text{for} \quad \kappa = ik_1;$$

$$w_1 = 0, \quad w_2 = i, \quad w_3 = -1, \quad \text{for} \quad \kappa = -ik_1.$$

Each satisfies the relation

$$w_1^2 + w_2^2 + w_3^2 = 0,$$

and each consists of a real part and an imaginary part of unit magnitude. The corresponding parts are mutually perpendicular. We also note that, considered as an operator, a skew-symmetric tensor of second order behaves like a vector product. More precisely, if the tensor operates on a vector $r_j$ then

$$u_i = k_{ij}r_j = -e_{ijm}k_m r_j = e_{mji}k_m r_j. \tag{4.49}$$

By using (4.22) one verifies that the result represents the vector product of the associated pseudo-vector $\mathbf{k}$ with the vector $\mathbf{r}$.

We have seen that the vector product of two polar vectors is a pseudo-vector; since one of the factors here is itself a pseudo-vector, the product becomes again a polar vector. This remark supplies the kinematic interpretation underlying the designation of the principal direction of the tensor as the axis of rotation, and of $\mathbf{k}$ as the pseudo-vector of rotation. If the vector $\mathbf{r}$ represents the position vector of a point in space with respect to a point on the axis of rotation, the vector $\mathbf{u}$ is the moment of the pseudo-vector $\mathbf{k}$ with respect to that point. The field of moments of $\mathbf{k}$ so generated may be identified with the displacement field of a solid in a finite rotation about the axis. This point of view will be developed during the kinematic analysis of finite displacements.

## 4.8. Matrix Representation of Tensor Operations

Tensors of order up to two may be represented by matrices, and the clarity of the matrix formulation may be advantageous in organizing the operations required to reach an intended goal. A scalar is still represented by its unique invariant symbol, a vector $u_i$ by a matrix $u$ with a single column, and a tensor $a_{ij}$ of order two by a square matrix $A$. The orthogonal matrix $T$ for a change of frame was introduced in Section 4.1.

Here, as with the explicit subscript notation, the representation pertains to a designated frame.

The laws (4.13) and (4.14) for tensor transformation of a polar vector have the respective matrix representations

$$n = T^T n' \quad \text{and} \quad n' = Tn.$$

For the transformation laws of a tensor of order two, one may either interpret the formulas (4.15) or use the invariance of a bilinear form

$$\phi = n^T A m = n'^T TAT^T m' = n'^T A' m'$$

to obtain the result

$$A' = TAT^T \quad \text{whence also} \quad A = T^T A' T.$$

While Kronecker's isotropic tensor is represented by the identity matrix $I$, the pseudo-tensor $e_{ijk}$ of order three has no matrix equivalent and should be replaced as an operator by an appropriate formalism. One may be based, as follows, on the vector product.

The skew matrix formed with an associated pseudo-vector $k$ is written

$$[k] = \begin{pmatrix} 0 & -k_3 & k_2 \\ k_3 & 0 & -k_1 \\ -k_2 & k_1 & 0 \end{pmatrix} = -[k]^T.$$

The vector product $k \times r$ is represented by

$$[k]r = -[r]k \quad \text{and we have} \quad [k]k = 0. \tag{4.50}$$

One verifies directly the important relation

$$[[k]r] = rk^T - kr^T$$

whence easily follows the formula for the double vector product:

$$[[k]r]u = (k^Tu)r - (r^Tu)k.$$

On the other hand, one verifies directly that

$$[k][r] = -(k^Tr)I + rk^T. \tag{4.51}$$

As an exercise in applying these formulas let us study again the structure of a skew tensor $[k]$. By writing

$$\omega = \sqrt{k^Tk}$$

we can reduce the analysis to that of a skew tensor $[k]$ with unit associated pseudo-vector $n$, where

$$[k] = \omega[n], \quad n^Tn = 1.$$

Let $u$ be another unit vector orthogonal to $n$, so

$$u^Tn = 0, \quad u^Tu = 1.$$

By operating on this unit vector with $[n]$ one forms

$$v = [n]u.$$

This new vector is also perpenducular to $n$ because, by (4.50),

$$n^T v = n^T [n] u = 0.$$

It is also perpendicular to $u$ because

$$u^T v = u^T [n] u = -u^T [n]^T u = -u^T [n] u = 0.$$

Here, the first equality is from the definition of $v$, the second because $[n]^T = -[n]$, and the third from the equality of a scalar and its transpose.

By (4.51), finally, the new vector is a unit vector:

$$v^T v = -u^T [n][n] u = -u^T (-I + nn^T) u = u^T u = 1.$$

We now perform a tensor transformation with the matrix $T^T$ defined by its columns as

$$T^T = (n \quad u \quad v).$$

We obtain first

$$[k] T^T = \omega([n]n \quad [n]u \quad [n]v) = \omega(0 \quad v \quad -u)$$

because

$$[n]v = [n]\ [n]u = (-I + nn^T)u = -u,$$

and finally

$$T[k]T^T = \omega \begin{pmatrix} 0 & n^T v & -n^T u \\ 0 & u^T v & -u^T u \\ 0 & v^T v & -v^T u \end{pmatrix} = \omega \begin{pmatrix} 0 & 0 & 0 \\ 0 & 0 & -1 \\ 0 & 1 & 0 \end{pmatrix}.$$

This is one of the canonical forms to which the tensor is reducible by a real orthogonal transformation.

For the unitary transformation

$$T^* = \left(n \quad \frac{u - iv}{\sqrt{2}} \quad \frac{v - iu}{\sqrt{2}}\right), \qquad T^* T = I,$$

where $T^*$ the transposed conjugate of $T$, one obtains a complex diagonal form:

$$T[k]T^* = \begin{pmatrix} 0 & 0 & 0 \\ 0 & i\omega & 0 \\ 0 & 0 & -i\omega \end{pmatrix}.$$

# Chapter 5
# The Equations of Linear Elasticity

## 5.1. Compatibility of Strains in a Simply Connected Domain

The geometric linearization

$$|D_i u_j| << 1 \tag{5.1}$$

allowed the displacement gradient tensor

$$D_i u_j = \varepsilon_{ij} + \omega_{ji} \tag{5.2}$$

to be expressed as the sum of a symmetric tensor of infinitesimal strains

$$\varepsilon_{ij} = \frac{1}{2}(D_i u_j + D_j u_i) = \varepsilon_{ji} \tag{5.3}$$

and a skew tensor of infinitesimal rotations

$$\omega_{ji} = \frac{1}{2}(D_i u_j - D_j u_i) = -\omega_{ij}. \tag{5.4}$$

If the latter expression is replaced by its associated pseudo-vector

$$\omega_m = \frac{1}{2} e_{mqj} D_q u_j \tag{5.5}$$

the decomposition of the gradient tensor becomes

$$D_i u_j = \varepsilon_{ij} + e_{ijm}\omega_m. \tag{5.6}$$

We shall now study the conditions which the rotation and strain fields must satisfy in order to be compatible.

There are nine equations in (5.6) and elimination of the three displacement components yields a set of necessary conditions for the rotations and strains. Formal elimination begins with the equations

$$e_{ipq}D_pD_iu_j = 0 \tag{5.7}$$

for each pair $(q,j)$, expressing simply the commutativity in the order of calculating the second partial derivatives of $u_j$. Substitution from (5.6) yields

$$e_{ipq}D_p\varepsilon_{ij} + e_{ipq}e_{ijm}D_p\omega_m = 0.$$

Now

$$e_{ipq}e_{ijm}D_p\omega_m = \binom{p \quad q}{j \quad m}D_p\omega_m = D_j\omega_q - \delta_{qj}D_m\omega_m,$$

and (5.5) shows that

$$D_m\omega_m = \frac{1}{2}\operatorname{div}\operatorname{rot} u = 0; \tag{5.8}$$

thus the result may be put into the form

$$D_j\omega_q = e_{piq}D_p\varepsilon_{ij}. \tag{5.9}$$

Equations (5.9) are known as *Beltrami's* equations. They express the gradient of the components of rotation in terms of the first derivatives of the strain field. On setting $j = q$ one sees that the rotations have the property (5.8) because of the symmetry of the strain tensor. The components of rotation may now be eliminated among the equations (5.9) by a repeating of the process, i.e., by noting that

$$e_{rjm}D_rD_j\omega_q = 0$$

for each pair (m,q), so that (5.9) leads to the equations

$$T_{qm} = 0, \tag{5.10}$$

where

$$T_{qm} = e_{piq}e_{rjm}D_pD_r\varepsilon_{ij} = T_{mq}. \tag{5.11}$$

$T_{qm}$ is called the tensor of incompatibility of the strains. Its symmetry appears clearly on exchange of the dummy indices $i$ and $j$ on one hand and $p$ and $r$ on the other. One sees also that for *any* $\varepsilon_{ij}$ (not necessarily a compatible strain tensor), the tensor defined by (5.11) satisfies the differential equations

$$D_qT_{qm} = 0. \tag{5.12}$$

That the incompatibility tensor should vanish at each point of the domain is seen to be a necessary condition for the existence of univalent (single-valued) rotation and displacement fields.

As we shall now see, this condition is also sufficient if the domain is simply connected. The proof uses Stokes' theorem

$$\oint \mathbf{a} \cdot dx = \int_S \mathbf{n} \cdot \operatorname{rot} \mathbf{a}\, dS.$$

The left member is the circulation of a vector field $\mathbf{a}$ around a closed contour of the domain. On the right is the flux of the curl of this vector across a surface of the domain bounded by the contour. If $\mathbf{a}$ is a polar vector, rot $\mathbf{a}$ is a pseudovector and the positive sense of the normal $\mathbf{n}$ is determined by the convention of pointing to the right if the reference frame is right-handed, but to the left if the frame is left-handed. In tensor notation

$$\oint a_j \, dx_j = \int_S n_r e_{rmj} D_m a_j \, dS. \tag{5.13}$$

The proof begins with the observation that in order to have a single-valued rotation vector at each point, it is necessary that on each closed circuit we have

$$\oint d\omega_q = \oint D_j \omega_q \, dx_j = 0. \tag{5.14}$$

If the domain is simply connected, then by definition each closed circuit may be contracted to a point without leaving the domain. During this contraction a surface is generated whose points all belong to the domain. Now, by using Stokes' theorem, the necessary conditions (5.14) take an equivalent form

$$\int_S n_r e_{rjm} D_m D_j \omega_q \, dS = 0$$

or, from Beltrami's equations,

$$\int_S n_r T_{rq} \, dS = 0. \tag{5.15}$$

If the incompatibility tensor vanishes in the entire domain then the conditions for the existence of a univalent rotation field are satisfied. We note that the rotation field so integrated is defined only up to a rotation vector

$$\omega_q = \beta_q = \text{constant},$$

because such a vector satisfies Beltrami's equations with vanishing right hand sides.

The second part of the proof entails constructing the displacement field by integrating equations (5.6). Here the conditions for a univalent field are that for every closed circuit

$$\oint du_j = \oint D_i u_j \, dx_i = 0.$$

After an application of Stokes' theorem they become

$$\int_S n_r e_{rpi} D_p D_i \, u_j \, dS = 0$$

and are satisfied exactly by the equation (5.7) from which Beltrami's equations were derived. The displacement field so constructed is therefore univalent, but it is defined only up to a field

$$u_j = a_j + e_{ijm} x_i \beta_m$$

where $a_j$, $\beta_m$ are six arbitrary constants. If the origin O of the Cartesian frame has been chosen within the domain, then the parameters $a_j$ and $\beta_m$ may be considered as the initial values of the displacements and rotations at O for the integration paths emanating from that point. We may therefore state the following theorem:

> In a simply connected domain, the vanishing at every point, including the boundary, of the six components of the incompatibility tensor is a necessary and sufficient condition for the existence of univalent rotation and displacement fields. The integration of a compatible strain field determines the displacements only up to an arbitrary infinitesimal rigid body displacement.

The complexity of the compatibility conditions (5.10) makes them difficult to use for constructing compatible strain fields, although one derives them very easily from the displacement field.

We now state and prove two lemmas starting with partial data on the strain field. We divide the components of the incompatibility tensor into two groups: the diagonal group

$$\begin{aligned} T_{11} &= D_2D_2\varepsilon_{33} + D_3D_3\varepsilon_{22} - 2\,D_2D_3\varepsilon_{23}, \\ T_{22} &= D_3D_3\varepsilon_{11} + D_1D_1\varepsilon_{33} - 2\,D_3D_1\varepsilon_{31}, \\ T_{33} &= D_1D_1\varepsilon_{22} + D_2D_2\varepsilon_{11} - 2\,D_1D_2\varepsilon_{12}, \end{aligned} \tag{5.16}$$

and the off-diagonal group

$$\begin{aligned} T_{23} &= -D_2D_3\varepsilon_{11} + D_1(-D_1\varepsilon_{23} + D_2\varepsilon_{31} + D_3\varepsilon_{12}), \\ T_{31} &= -D_3D_1\varepsilon_{22} + D_2(-D_2\varepsilon_{31} + D_3\varepsilon_{12} + D_1\varepsilon_{23}), \\ T_{12} &= -D_1D_2\varepsilon_{33} + D_3(-D_3\varepsilon_{12} + D_1\varepsilon_{23} + D_2\varepsilon_{31}). \end{aligned} \tag{5.17}$$

Lemma 1. An arbitrary sufficiently differentiable field of shear strains $(2\varepsilon_{23}, 2\varepsilon_{31}, 2\varepsilon_{12})$ may be completed by a field $(\varepsilon_{11}, \varepsilon_{22}, \varepsilon_{33})$ of specific elongations in such a way as to be compatible.

In other words, it is possible to construct a displacement field such that

$$\begin{gathered} D_2u_3 + D_3u_2 = 2\varepsilon_{23}, \qquad D_3u_1 + D_1u_3 = 2\varepsilon_{31}, \\ D_1u_2 + D_2u_1 = 2\varepsilon_{12} \end{gathered} \tag{5.18}$$

where the right hand sides are given.

We move directly to integrating the field of displacements, starting with the data on shear strains. Then we obtain the complementary specific elongations from the equations

$$\varepsilon_{11} = D_1u_1, \quad \varepsilon_{22} = D_2u_2, \quad \varepsilon_{33} = D_3u_3, \tag{5.19}$$

substitute them into the compatibility conditions

$$T_{23} = 0, \quad T_{31} = 0, \quad T_{12} = 0,$$

suppress the differentiation operator common to all the terms, and obtain separate equations for each component of the displacement

$$\begin{aligned} D_2D_3u_1 &= -D_1\varepsilon_{23} + D_2\varepsilon_{31} + D_3\varepsilon_{12} \\ D_3D_1u_2 &= -D_2\varepsilon_{31} + D_3\varepsilon_{12} + D_1\varepsilon_{23} \\ D_1D_2u_3 &= -D_3\varepsilon_{12} + D_2\varepsilon_{23} + D_2\varepsilon_{31}. \end{aligned} \tag{5.20}$$

These are also the equations yielded by eliminating two of the displacements from equations (5.18).

Each particular solution of these equations also satisfies the equations

$$\begin{aligned} D_3[2\varepsilon_{12} - D_2u_1 - D_1u_2] &= 0, \\ D_1[2\varepsilon_{23} - D_3u_2 - D_2u_3] &= 0, \\ D_2[2\varepsilon_{31} - D_1u_3 - D_3u_1] &= 0, \end{aligned}$$

obtained by simply adding equations (5.20) two by two; for this particular solution one therefore has the relations

$$\begin{aligned} 2\varepsilon_{12} &= D_2u_1 + D_1u_2 + E_{12}(x_1,x_2), \\ 2\varepsilon_{23} &= D_3u_2 + D_2u_3 + E_{23}(x_2,x_3), \\ 2\varepsilon_{31} &= D_1u_3 + D_3u_1 + E_{31}(x_3,x_1), \end{aligned} \tag{5.21}$$

which differs from the required equations (5.18) only by the presence of the arbitrary functions $E_{ij}$.

At the same time, equations (5.20) themselves show that from each particular solution one may subtract

$$u_1 = F_{12}(x_1,x_2) + F_{13}(x_1,x_3)$$
$$u_2 = F_{21}(x_2,x_1) + F_{23}(x_2,x_3)$$
$$u_3 = F_{31}(x_3,x_1) + F_{32}(x_3,x_2)$$

and it is obviously always possible to choose the arbitrary functions $E_{ij}$ in such a way that

$$D_2F_{12} + D_1F_{21} = E_{12}$$
$$D_3F_{23} + D_2F_{32} = E_{23}$$
$$D_1F_{31} + D_3F_{13} = E_{31}.$$

In other words, no generality is lost in making $E_{ij} = 0$ in equations (5.21). They and (5.19) complete the proof of the Lemma.*

Lemma 2. An arbitrary sufficiently differentiable field of specific elongations $(\varepsilon_{11}, \varepsilon_{22}, \varepsilon_{33})$ may be completed by a field of shear strains so as to be compatible.

Here it is enough to integrate the displacements by equations (5.19) and to define the complementary field by equations (5.18).

Applied in a different context and connected with properties (5.12), these lemmas allow recovery of the following results due to *Washizu:*

Theorem 1. If a strain field satisfies the compatibility conditions $T_{11} = 0$, $T_{22} = 0$, and $T_{33} = 0$ in a region $V$ and on its boundary $\partial V$, then it is sufficient for it to satisfy the complementary conditions $T_{23} = 0$, $T_{31} = 0$, and

*Editor's note: The author's reasoning here is somewhat unclear.

$T_{12} = 0$ on $\partial V$ in order for it to satisfy them also in $V$.

According to Lemma 1, for any sufficiently differentiable $T_{23}$, $T_{31}$, and $T_{12}$ in $V$ there is a vector field $U$ such that

$$T_{23} = D_2U_3 + D_3U_2, \quad T_{31} = D_3U_1 + D_1U_3, \quad T_{12} = D_1U_2 + D_2U_1.$$

Now

$$\begin{aligned}\int_V (T_{23}^2+T_{31}^2+T_{12}^2)\,dV &= \int_V \{T_{23}(D_2U_3+D_3U_2) + T_{31}(D_3U_1+D_1U_3) \\ &\qquad + T_{12}(D_1U_2+D_2U_1)\}\,dV \\ &= \int_{\partial V} [U_1(n_2T_{21}+n_3T_{31}) + U_2(n_3T_{32}+n_1T_{12}) + U_3(n_1T_{13}+n_2T_{23})]\,dS \\ &\quad - \int_V [U_1(D_2T_{21}+D_3T_{31}) + U_2(D_3T_{32}+D_1T_{12}) \\ &\qquad + U_3(D_1T_{13}+D_2T_{23})]\,dV.\end{aligned}$$

By properties (5.12) and our hypotheses we have

$$D_2T_{21} + D_3T_{31} = -D_1T_{11} = 0,$$
$$D_3T_{32} + D_1T_{12} = -D_2T_{22} = 0,$$
$$D_1T_{13} + D_2T_{23} = -D_3T_{33} = 0,$$

while the boundary integral also vanishes by the hypotheses. Therefore

$$\int_V (T_{23}^2 + T_{31}^2 + T_{12}^2)\,dV = 0$$

and the stated result follows.

Theorem 2. If a strain field satisfies the compatibility conditions $T_{23} = 0$, $T_{31} = 0$, and $T_{12} = 0$ in a region $V$ and on its boundary $\partial V$, then it is sufficient for it to satisfy the complementary conditions $T_{11} = 0$, $T_{22} = 0$, and $T_{33} = 0$ on $\partial V$ in order for it to satisfy them also in $V$.

The proof is similar but uses Lemma 2, which permits writing

$$T_{11} = D_1U_1, \quad T_{22} = D_2U_2, \quad T_{33} = D_3U_3.$$

We define a *regular* strain field as one which has these properties:

1) The strains are continuous along with their first and second partial derivatives. This is *V. Volterra's* regularity condition.

2) The incompatibility tensor formed with the second derivatives vanishes at every point of the field.

The first requirement may occasionally be relaxed for the continuity of the second derivatives on certain singular surfaces.

We may summarize the principal result of this section as follows:

In a simply connected region, every regular strain field may be induced by a change of configuration.

We shall see in the next section that this proposition, in order to hold in a multiply connected region, demands further properties of the strain field.

## 5.2. Compatibility of Strains in a Multiply Connected Region

We consider a region homeomorphic to a torus and study the nature of the closed curves which may be drawn in it. We call two circuits (closed curves) *reconcilable* if they can be made to coincide by a continuous deformation without leaving

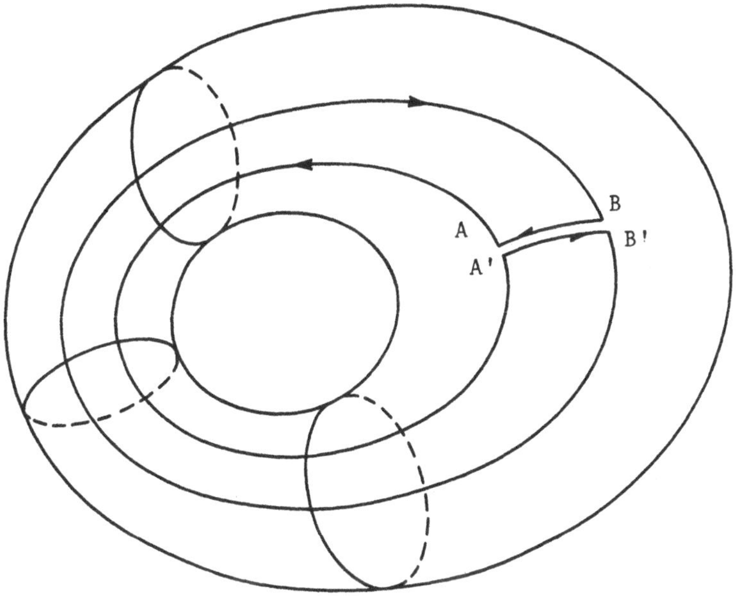


Figure 5.1

the region. We call a circuit *reducible* if it is reconcilable with a single point. All reducible circuits are reconcilable and therefore belong to an equivalence class. The region also has irreducible circuits, such as AA'A or BB'B; they are not reconcilable with the reducible circuits, but are reconcilable with each other; they form a second equivalence class. Such a region, having two equivalence classes of closed circuits is, by definition, *doubly connected.*

The region shown below, homeomorphic to a double torus, is *triply connected.* Its three equivalence classes are formed by reducible closed circuits such as the circuit (1), and by irreducible circuits of types (2) and (3). A circuit of type (4), drawn as a dashed line, is reconcilable with a pair of

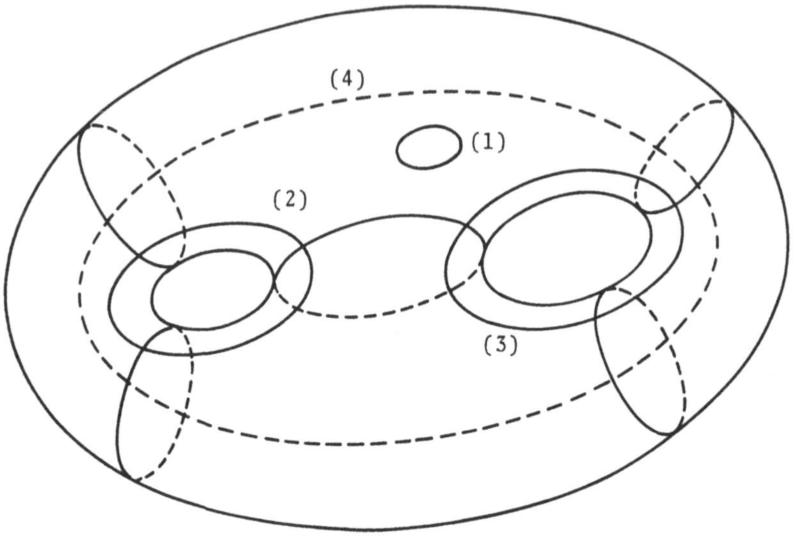


Figure 5.2

circuits, one each of types (2) and (3), and is not regarded as an element of a new equivalence class.

We return to the doubly connected region and join the two irreducible circuits AA'A and BB'B by a "bridge" consisting of two arcs AA' and BB' which will be made to approach each other in the limit. The composite arc AA'B'BA is reducible; it has a supporting surface entirely contained in the region. As a consequence, by applying Stokes' theorem as in the preceding section, one may conclude that every regular strain field integrated by equations (5.9) and then (5.6), generates a field of rotations and displacements with the properties

$$\int_{AA'} d\omega_q + \int_{A'B'} d\omega_q + \int_{B'B} d\omega_q + \int_{BA} d\omega_q = 0, \tag{5.22}$$

$$\int_{AA'} du_j + \int_{A'B'} du_j + \int_{B'B} du_j + \int_{BA} du_j = 0. \tag{5.23}$$

We now study the first of these properties of univalence in the extreme case in which A'B' = AB. The first integral becomes that on an irreducible circuit passing through A and equals the jump in the value of $\omega_q$ when the circuit is traversed once in the direction indicated:

$$\int_{AA'} d\omega_q = \oint_A d\omega_q = \Delta_A \omega_q .$$

Similarly, the third integral becomes

$$\int_{B'B} d\omega_q = -\oint_B d\omega_q = -\Delta_B \omega_q ,$$

because the irreducible circuit passing through B is traversed in the opposite direction. By using (5.6) the integrals on the bridge may be expressed in terms of the strains:

$$\int_{A'B'} d\omega_q = \int_{AB} D_j \omega_q \, dx_j = \int_{AB} e_{piq} D_j \varepsilon_{ij} \, dx_j$$

$$\int_{BA} d\omega_q = -\int_{AB} D_j \omega_q \, dx_j = -\int_{AB} e_{piq} D_j \varepsilon_{ij} \, dx_j .$$

They mutually cancel because they traverse the same arc in opposite directions. Equation (5.22) is thus equivalent with

$$\Delta_A \omega_q = \Delta_B \omega_q = \Delta \omega_q . \tag{5.24}$$

The increment in the rotation vector is the same for all irreducible circuits in the same equivalence class traversed once in the same direction. We say that the components $\omega_q$ are *polydromes* with *cyclic constants* $\Delta\omega_q$.

In the same limiting case the property of univalence (5.23) generates a slightly more complicated property of the displacement field. One has again

$$\int_{AA'} du_j = \oint_A du_j = \Delta_A u_j,$$

$$\int_{B'B} du_j = -\oint_B du_j = -\Delta_B u_j.$$

But the integrals on the bridge do not disappear completely. From (5.6) there follows

$$\begin{aligned}\int_{A'B'} D_i u_j dx_i &+ \int_{BA} D_i u_j dx_i \\ &= \int_{AB} [\varepsilon_{ij} + e_{ijm}(\omega_m + \Delta\omega_m)] dx_i - \int_{AB} (\varepsilon_{ij} + e_{ijm}\omega_m) dx_i \\ &= e_{ijm}\Delta\omega_m \int_{AB} dx_i,\end{aligned}$$

where we have used the continuity of the strains and the cyclic constants (5.24) for the multi-valued rotations. The final result is

$$\Delta_B u_j = \Delta_A u_j + e_{ijm}\Delta\omega_m(x_{iB} - x_{iA}). \tag{5.25}$$

In summary, we have shown that polydromy of the displacement field for a regular strain may be defined by six parameters $(\Delta\omega_m, \Delta_A u_j)$, with formula (5.25) then allowing the calculation of the cyclic constants of the displacement for a different point B.

In a multiply connected region every equivalence class of irreducible circuits has its own system of parameters of polydromy. We state a few consequences of that fact.

First, the necessary and sufficient conditions that a strain field in a multiply connected region be the result of a change of configuration are: (a) that the field be regular (i.e., compatible and sufficiently differentiable), and (b) that the conditions for univalence

$$\Delta\omega_p = \oint e_{piq} D_q \varepsilon_{ij}\, dx_j = 0, \tag{5.26}$$

$$\Delta u_j = \oint (\varepsilon_{ij} + e_{ijm}\omega_m)dx_i = 0 \qquad (5.27)$$

be satisfied for at least one circuit in each irreducible equivalence class.

The necessity follows at once from the fact that associated with every change of configuration there is a univalent field of rotations and displacements. Conditions (a) and (b) will be sufficient if they imply (5.26) and (5.27) for an arbitrary closed circuit of the region. We saw in the preceding section that the regularity of the field implies the validity of (5.26) and (5.27) for every reducible circuit. At the same time, (5.24) shows that if condition (5.26) holds for one circuit then it holds for every other circuit in the same equivalence class; in this case (5.25) reduces to

$$\Delta_B u_j = \Delta_A u_j = \Delta u_j$$

and the same result may be obtained for condition (5.27).

In contrast to the compatibility conditions (5.10) which may be described as "local", the further conditions (5.26) and (5.27) are "global" compatibility conditions. One possible interpretation for the formation of regular fields *not* satisfying the global compatibility conditions is supplied by the idea of dislocation due to *Weingarten* and *Volterra*. Every multiply connected domain may be reduced to simple connectivity by introducing, for each equivalence class of irreducible circuits, a *cut* which no circuit is permitted to cross. Figure 5.3 shows a cut capable of preventing the closure of every irreducible circuit in the doubly connected region previously considered. Each cut has two faces, and any two originally irreducible circuits of the same class (e.g.,

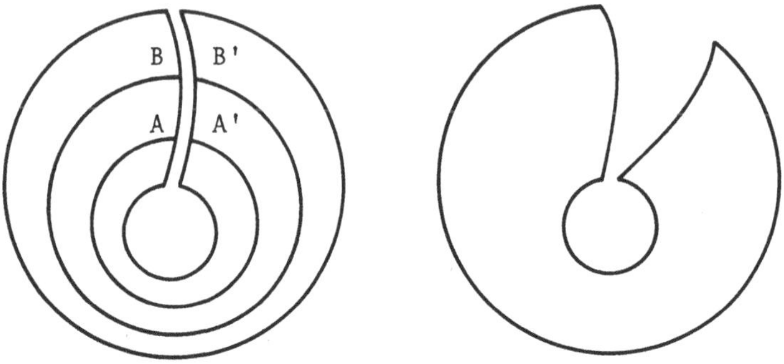


Figure 5.3

AA', BB'), may be bridged by a pair of segments (AB, A'B'), one on each face of the cut, to form a reducible circuit.

Now assume the existence of a regular strain field in the region. Even if it does not satisfy the global compatibility conditions it defines (up to an infinitesimal rigid displacement) a univalent displacement for the cut region, which is simply connected. Thus it corresponds to a change of configuration of this region. It is therefore possible to annul all the strains by imposing on each point the negative of its original displacement vector. One may now interpret formulas (5.24) and (5.25) as the negatives of the relative motion which would arise during this process of complete strain relaxation. If A is an arbitrary reference point on the cut and B any other point on the same cut, formula (5.25) shows that the relative displacement between the two edges has the form of an infinitesimal rigid body displacement. Volterra called this a *distortion* but in modern terminology it is a *dislocation*.

Starting with an initially dislocated region, it is

possible by closing up the cuts to generate a strain field which is regular but violates the conditions of global compatibility. One sees that this operation causes no singularity in the strain field which would permit the identification of the location of the original cuts.

A dislocation constitutes one possible process by which an elastic medium may develop initial stresses which cannot be fully relaxed by a change of configuration.

## 5.3. Principal Elongations and Fundamental Invariants of Strain

Being symmetric, the infinitesimal strain tensor has at each point principal directions corresponding to eigenvalues which are here the *principal strains* denoted by $\varepsilon_i$. The corresponding fundamental invariants are given by the formulas

$$\begin{aligned} I_1 &= \varepsilon_{ii} = \varepsilon_1 + \varepsilon_2 + \varepsilon_3 = D_i u_i = \operatorname{div} u, \\ I_2 &= \tfrac{1}{2} e_{ijk} e_{mnk} \varepsilon_{im} \varepsilon_{jn} = \tfrac{1}{2}[\varepsilon_{ii}\varepsilon_{jj} - \varepsilon_{ij}\varepsilon_{ji}] \\ &\qquad = \varepsilon_1\varepsilon_2 + \varepsilon_2\varepsilon_3 + \varepsilon_3\varepsilon_1, \\ I_3 &= \det\{\varepsilon_{ij}\} = \varepsilon_1\varepsilon_2\varepsilon_3. \end{aligned} \tag{5.28}$$

One sees that, up to terms of the first order in displacement gradients,

$$\begin{aligned} \det J &= e_{ijk}(\delta_{1i} + \varepsilon_{1i})(\delta_{2j} + \varepsilon_{2j})(\delta_{3k} + \varepsilon_{3k}) \\ &= e_{123} + \varepsilon_{11} + \varepsilon_{22} + \varepsilon_{33} = 1 + I_1, \end{aligned} \tag{5.29}$$

and that the first fundamental invariant therefore represents the specific increase in volume. Also, according to the general theory, the second invariant $\Gamma_2$ of the strain deviator

$$\hat{\varepsilon}_{ij} = \varepsilon_{ij} - \frac{1}{3} I_1 \delta_{ij} \tag{5.30}$$

has the value

$$\Gamma_2 = -\frac{1}{2}\hat{\varepsilon}_{ij}\hat{\varepsilon}_{ji} = -\frac{1}{6}\{(\varepsilon_{11}-\varepsilon_{22})^2 + (\varepsilon_{22}-\varepsilon_{33})^2 + (\varepsilon_{33}-\varepsilon_{11})^2 + 6(\varepsilon_{23}^2+\varepsilon_{31}^2+\varepsilon_{12}^2)\}. \tag{5.31}$$

## 5.4. Principal Stresses and Fundamental Invariants of the Stress State

We shall continue to use the notation $\tau_{ij}$ to represent the true stress, which is symmetric. However, as we have seen, when the changes of configuration are infinitesimal there is practically no difference between the true stress and Piola's stress. At each point the tensor $\tau_{ij}$ has at least one system of principal directions associated with eigenvalues which we refer to as *principal stresses* and denote by $\sigma_i$. The fundamental invariants are

$$J_1 = \tau_{ii} = \sigma_1 + \sigma_2 + \sigma_3,$$

$$J_2 = \frac{1}{2}e_{ijk}e_{mnk}\tau_{im}\tau_{jn} = \frac{1}{2}[\tau_{ii}\tau_{jj} - \tau_{ij}\tau_{ji}] = \sigma_1\sigma_2 + \sigma_2\sigma_3 + \sigma_3\sigma_1,$$

$$J_3 = \det\{\tau_{ij}\} = \sigma_1\sigma_2\sigma_3.$$

The second fundamental invariant of the deviatoric tensor

$$\hat{\tau}_{ij} = \tau_{ij} - \frac{1}{3} J_1 \delta_{ij} \tag{5.33}$$

will be denoted by

$$\Sigma_2 = -\frac{1}{2}\hat{\tau}_{ij}\hat{\tau}_{ji} = -\frac{1}{6}\{(\tau_{11}-\tau_{22})^2 + (\tau_{22}-\tau_{33})^2 + (\tau_{33}-\tau_{11})^2 + 6(\tau_{23}^2+\tau_{31}^2+\tau_{12}^2)\}. \tag{5.33'}$$

## 5.5. Octahedral Stresses and Strains

Recall Cauchy's formulas (2.3) which give the surface traction vector as a function of the stress tensor and the orientation of the normal to the surface. With the deformed configuration as reference, and consequently Piola's stress equivalent to true stress,

$$t_j = n_i \tau_{ij}. \tag{5.34}$$

Suppose now that the frame is oriented according to a system of principal directions, so that the stress tensor has a diagonal matrix

$$\begin{pmatrix} \sigma_1 & 0 & 0 \\ 0 & \sigma_2 & 0 \\ 0 & 0 & \sigma_3 \end{pmatrix}.$$

Then (5.34) reduces to

$$t_1 = n_1\sigma_1, \quad t_2 = n_2\sigma_2, \quad t_3 = n_3\sigma_3. \tag{5.35}$$

The traction normal to a face is

$$t_n = n_i t_i = n_1^2\sigma_1 + n_2^2\sigma_2 + n_3^2\sigma_3 \tag{5.36}$$

so that, on each of the eight faces of an octahedron, equally inclined to the principal axes $(n_i = \pm 1/\sqrt{3})$, the normal traction has the same value, viz.,

$$t_{on} = \tfrac{1}{3}(\sigma_1 + \sigma_2 + \sigma_3) = \tfrac{1}{3}J_1. \tag{5.37}$$

One may evaluate the magnitude of the shear traction on any inclined face by the Pythagoras' theorem; thus

$$t_t^2 = t_1^2 + t_2^2 + t_3^2 - t_n^2 = n_1^2n_2^2(\sigma_1-\sigma_2)^2 + n_2^2n_3^2(\sigma_2-\sigma_3)^2 + n_3^2n_1^2(\sigma_3-\sigma_1)^2. \tag{5.38}$$

For the faces of the octahedron this gives

$$(3t_{ot})^2 = (\sigma_1-\sigma_2)^2 + (\sigma_2-\sigma_3)^2 + (\sigma_3-\sigma_1)^2 = -6\Sigma_2. \quad (5.39)$$

The *octahedral normal* and *shear stresses* are thus related respectively to the first invariant of the stress state and the second invariant of its deviator.

A similar calculation may obviously be made for the strain tensor. It leads to defining an octrahedral elongation

$$\varepsilon_{on} = \frac{1}{3}(\varepsilon_1 + \varepsilon_2 + \varepsilon_3) = \frac{1}{3} I_1 \quad (5.40)$$

and an octahedral shear strain $\gamma_{ot} = 2\varepsilon_{ot}$ such that

$$(3\varepsilon_{ot})^2 = (\varepsilon_1-\varepsilon_2) + (\varepsilon_2-\varepsilon_3)^2 + (\varepsilon_3-\varepsilon_1)^2 = -6\Gamma_2. \quad (5.41)$$

In general, the principal directions of strain and stress at the same point will coincide only accidentally. By one possible definition, an *isotropic* medium is one whose constitutive equations require that the principal directions of stress and strain always coincide.

## 5.6. Mohr's Circles

If the octahedral quantities illustrate certain fundamental invariants, *Mohr's circles* illustrate graphically the tensorial character of symmetric, second order tensors in three-dimensional Euclidean space.

Let us eliminate $n_2^2$ and $n_3^2$ between the relation (5.36), rewritten here as

$$t_n - n_1^2\sigma_1 = n_2^2\sigma_2 + n_3^2\sigma_3,$$

the relation (5.38) in the form

$$t_t^2 + t_n^2 - n_1^2\sigma_1^2 = n_2^2\sigma_2^2 + n_3^2\sigma_3^2,$$

and the relation between direction cosines

$$1 - n_1^2 = n_2^2 + n_3^2.$$

The result is

$$\det \begin{vmatrix} t_n - n_1^2\sigma_1 & \sigma_2 & \sigma_3 \\ t_t^2 + t_n^2 - n_1^2\sigma_1^2 & \sigma_2^2 & \sigma_3^2 \\ 1 - n_1^2 & 1 & 1 \end{vmatrix} = 0$$

or, after reduction,

$$t_t^2 + \left(t_n - \frac{\sigma_2+\sigma_3}{2}\right)^2 = \left(\frac{\sigma_2-\sigma_3}{2}\right)^2 + n_1^2(\sigma_1-\sigma_2)(\sigma_1-\sigma_3). \qquad (5.42)$$

In a $(t_n, t_t)$ plane the locus of this equation depends on the parameter $n_1^2$. If we keep $n_1^2$ constant, the normal is confined to a cone making a constant angle with the first principal direction (in both orientations). To expedite the discussion it is necessary to impose the restriction

$$\sigma_1 \geq \sigma_2 \geq \sigma_3, \qquad (5.43)$$

a situation which may always be brought about by modifying the order of the principal directions.

The locus of (5.42) is a circle with center at

$$\left(\frac{\sigma_2 + \sigma_3}{2}, 0\right)$$

and squared radius given by the right hand side of (5.42). For $n_1 = 0$, the normal is confined to the plane perpendicular to the first principal direction, and we have the *Mohr's circle* for $n_1 = 0$. From (5.43) we see that the circle arising from any nonzero value of $n_1^2$ has a larger radius such as, for example, that passing through the point P in the diagram below. Admissible points $(t_n, t_t)$ are confined to

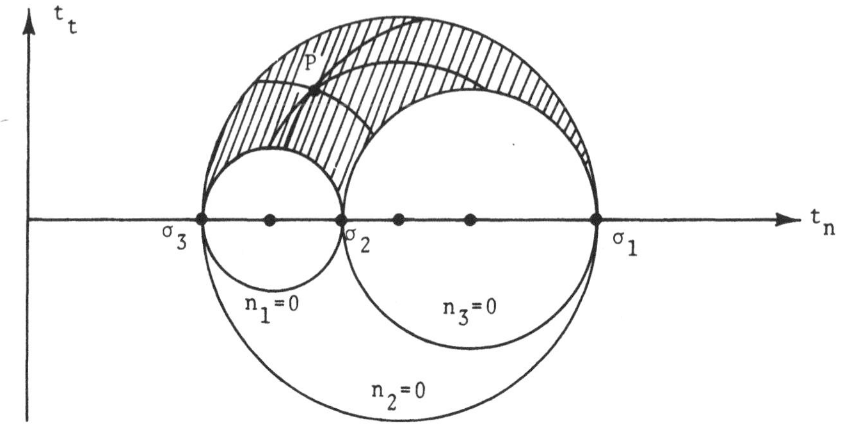


Figure 5.4

an arc at whose ends either $n_2$ or $n_3$ vanishes before becoming imaginary. When $n_1^2$ reaches its maximum of 1, the arc collapses to a single point $t_n = \sigma_1$, $t_t = 0$.

Similar elimination of $n_3^2$ and $n_1^2$ yields an analogous equation

$$t_t^2 + (t_n - \frac{\sigma_3+\sigma_1}{2})^2 = (\frac{\sigma_3-\sigma_1}{2})^2 + n_2^2(\sigma_2-\sigma_3)(\sigma_2-\sigma_1)$$

which arises directly from (5.42) by cyclic permutation of indices. It corresponds to a conical sweep of the normal about the second principal direction. This time, for $n_2^2 > 0$ the radius of the circle is smaller than that of Mohr's circle for $n_2 = 0$ and the limiting point for $n_2^2 = 1$ is $(\sigma_2, 0)$.

Finally, for the conical sweep about the third principal direction,

$$t_t^2 + (t_n - \frac{\sigma_1+\sigma_2}{2})^2 = (\frac{\sigma_1-\sigma_2}{2})^2 + n_3^2(\sigma_3-\sigma_1)(\sigma_3-\sigma_2)$$

and the radius for $n_3^2 > 0$ is again larger than that of Mohr's circle for $n_3 = 0$. The limit point for $n_3^2 = 1$ is

$(\sigma_3, 0)$.

The shaded area of Figure 5.4 shows which combinations of normal and tangential traction are possible. It also shows the existence of three relative maxima of the tangential traction, viz.,

$$\frac{\sigma_2 - \sigma_3}{2}, \quad \frac{\sigma_1 - \sigma_2}{2}, \quad \frac{\sigma_1 - \sigma_3}{2},$$

the last being the absolute maximum in this case. The three values correspond respectively to the orientations

$$n_1 = 0, \quad n_2^2 = n_3^2 = \frac{1}{2};$$

$$n_2 = 0, \quad n_3^2 = n_1^2 = \frac{1}{2};$$

$$n_3 = 0, \quad n_1^2 = n_2^2 = \frac{1}{2}.$$

## 5.7. Statics and Virtual Work

The symmetry

$$\tau_{ij} = \tau_{ji} \tag{5.44}$$

of the stress tensor is, as we have seen, the expression (exact for true stress and approximate for Piola's stress) of rotational equilibrium. The equations for translational equilibrium may be written

$$L_j(\tau_{mn}) = -D_i\tau_{ij} = \rho_o\, g_j, \tag{5.45}$$

the first equality defining the differential operators $L_j$ which act on the stresses to balance the body forces. A duality between statics and kinematics results from the "adjoint" character of the differential operators $L_j$ and those defined by the first equality in

$$M_{ij}(u_m) = \frac{1}{2}(D_iu_j + D_ju_i) = \varepsilon_{ij},$$

which operate on the displacements to yield the strains. Indeed, in view of (5.44) the expression

$$\tau_{ij}M_{ij}(u_m) - u_jL_j(\tau_{mn}) = \frac{1}{2}\tau_{ij}(D_iu_j + D_ju_i) + u_jD_i\tau_{ij}$$

may be converted to a divergence

$$\tau_{ij}D_iu_j + u_jD_i\tau_{ij} = D_i(u_j\tau_{ij}).$$

By integrating over the volume, and using the divergence theorem and Cauchy's equation for tractions

$$n_i\tau_{ij} = t_j,$$

one recovers the theorem of virtual work

$$\int_V(\tau_{ij}\varepsilon_{ij} - \rho_0 u_j g_j)dV = \int_{\partial V} u_j n_i \tau_{ij}\, dS = \int_{\partial V} t_j u_j\, dS. \tag{5.46}$$

One may therefore say that $u_j$ and $\tau_{ij}$ are adjoint variables, and that $\tau_{ij}$ and $\varepsilon_{ij}$ are congugate variables.

## 5.8. Taylor's Development of the Strain Energy

Provided that stress is a function of strain, the strain energy density is a function of the components of strain such that in an arbitrary local increment of strain $\delta\varepsilon_{ij}$ the increment of strain energy density $\delta W$ is

$$\begin{aligned}\delta W = \tau_{ij}\delta\varepsilon_{ij} = \tau_{11}\delta\varepsilon_{11} + \tau_{22}\delta\varepsilon_{22} + \tau_3\delta\varepsilon_{33} \\ + \tau_{23}\delta\gamma_{23} + \tau_{31}\delta\gamma_{31} + \tau_{12}\delta\gamma_{12},\end{aligned} \tag{5.47}$$

where, we recall, the shear strains are defined by

$$\gamma_{23} = 2\varepsilon_{23}, \quad \gamma_{31} = 2\varepsilon_{31}, \quad \gamma_{12} = 2\varepsilon_{12}.$$

Because $W$ is a perfect differential, the constitutive equations are expressible in a general way as

$$\tau_{11} = \frac{\partial W}{\partial \varepsilon_{11}}, \quad \tau_{22} = \frac{\partial W}{\partial \varepsilon_{22}}, \quad \tau_{33} = \frac{\partial W}{\partial \varepsilon_{33}}$$

$$\tau_{23} = \frac{\partial W}{\partial \gamma_{23}}, \quad \tau_{31} = \frac{\partial W}{\partial \gamma_{31}}, \quad \tau_{12} = \frac{\partial W}{\partial \gamma_{12}}. \tag{5.48}$$

If $W$ is considered as a function of $\varepsilon_{11}$, $\varepsilon_{22}$, $\varepsilon_{33}$, and of

$$\gamma_{23} = \varepsilon_{23} + \varepsilon_{32}, \quad \gamma_{31} = \varepsilon_{31} + \varepsilon_{13}, \quad \gamma_{12} = \varepsilon_{12} + \varepsilon_{21},$$

then the formal distinction thus introduced between $\varepsilon_{ij}$ and $\varepsilon_{ji}$ for $i \neq j$ allows the collection of the six formulas (5.48) into the single one

$$\tau_{ij} = \frac{\partial W}{\partial \varepsilon_{ij}},$$

which also automatically expresses the symmetry $\tau_{ji} = \tau_{ij}$.

Assume now that $W$ is a sufficiently regular function of its arguments to allow its development in a Taylor series. It will be sufficient to limit the development to terms of degree at most two:

$$W = W_o + \varepsilon_{ij}\left(\frac{\partial W}{\partial \varepsilon_{ij}}\right)_o + \frac{1}{2}\varepsilon_{ij}\varepsilon_{pq}\left(\frac{\partial^2 W}{\partial \varepsilon_{ij}\partial \varepsilon_{pq}}\right)_o. \tag{5.50}$$

$W_o$ is the value of the energy in the reference state, which state is chosen as $\varepsilon_{ij} = 0$. Without loss of generality $W_o$ may be taken as zero, i.e., the reference level of energy is taken to vanish.

The index zero attached to the partial derivatives indicates that they also are evaluated at the reference level. At the same time, because of (5.49), one may write

$$\left(\frac{\partial W}{\partial \varepsilon_{ij}}\right)_o = \tau^o_{ij}. \tag{5.51}$$

These are the initial stresses which may occur in the reference state and whose detailed experimental evaluation presents a difficult problem.

Finally, with $W_0 = 0$ and the notation

$$\left(\frac{\partial^2 W}{\partial \varepsilon_{ij} \partial \varepsilon_{pq}}\right)_0 = C_{ij}^{pq} \tag{5.52}$$

we may present Taylor's series in the form

$$W = \tau_{ij}^o \varepsilon_{ij} + \frac{1}{2} C_{ij}^{pq} \varepsilon_{ij} \varepsilon_{pq}, \tag{5.53}$$

while the relations (5.49) furnish the explicit constitutive equations

$$\tau_{ij} - \tau_{ij}^o = C_{ij}^{pq} \varepsilon_{pq}. \tag{5.54}$$

The limitation of the expansion to terms of the second degree thus corresponds to assuming a constitutive linearity between stresses and strains.

The *linear theory of elasticity* is characterized by simultaneous geometric and constitutive linearization.

One sees that (5.53) may also be written

$$W = \tau_{ij}^o \varepsilon_{ij} + \frac{1}{2} \varepsilon_{ij}(\tau_{ij} - \tau_{ij}^o) = \frac{1}{2} \varepsilon_{ij}(\tau_{ij} + \tau_{ij}^o). \tag{5.55}$$

This result, a consequence of the linearity of the constitutive equations, is the *Interior theorem of Clapeyron*. It is usually presented in the special case in which one assumes a *natural* reference state, i.e., one in which the initial stresses vanish.

The coefficients (5.52) are called *moduli of elasticity*. They form a Cartesian tensor of order four, with $3^4 = 81$ components. Involving second partial derivatives with respect to the components of a symmetric tensor, they inherit the following symmetry properties:

$C_{ij}^{pq} = C_{ji}^{pq}$ symmetry with respect to the lower indices

$C_{ij}^{pq} = C_{ij}^{qp}$ symmetry with respect to the upper indices

$C_{ij}^{pq} = C_{pq}^{ij}$ symmetry with respect to the vertical pairs of indices. (5.56)

The number of independent components is thus reduced to 21, as is confirmed directly by formula (5.53), where they have the status of coefficients of a quadratic form in six independent variables.

When studying the properties of isotropy in various elastic media we shall find it convenient to characterize the medium by a 6 × 6 matrix of coefficients of the linear system (5.54). If we perform the summations, then using the symmetries we obtain

$$\tau_{ij} - \tau_{ij}^{o} = C_{ij}^{11}\varepsilon_{11} + C_{ij}^{22}\varepsilon_{22} + C_{ij}^{33}\varepsilon_{33} + C_{ij}^{23}(\varepsilon_{23} + \varepsilon_{32}) + C_{ij}^{31}(\varepsilon_{31} + \varepsilon_{13}) + C_{ij}^{12}(\varepsilon_{12} + \varepsilon_{21}),$$

so that in matrix form

$$\tau = C\,\epsilon \qquad (5.57)$$

or

$$\begin{pmatrix} \tau_{11} \\ \tau_{22} \\ \tau_{33} \\ \tau_{23} \\ \tau_{31} \\ \tau_{12} \end{pmatrix} = \begin{pmatrix} C_{11}^{11} & C_{11}^{22} & C_{11}^{33} & C_{11}^{23} & C_{11}^{31} & C_{11}^{12} \\ C_{22}^{11} & C_{22}^{22} & C_{22}^{33} & C_{22}^{23} & C_{22}^{31} & C_{22}^{12} \\ C_{33}^{11} & C_{33}^{22} & C_{33}^{33} & C_{33}^{23} & C_{33}^{31} & C_{33}^{12} \\ C_{23}^{11} & C_{23}^{22} & C_{23}^{33} & C_{23}^{23} & C_{23}^{31} & C_{23}^{12} \\ C_{31}^{11} & C_{31}^{22} & C_{31}^{33} & C_{31}^{23} & C_{31}^{31} & C_{31}^{12} \\ C_{12}^{11} & C_{12}^{22} & C_{12}^{33} & C_{12}^{23} & C_{12}^{31} & C_{12}^{12} \end{pmatrix} \begin{pmatrix} \varepsilon_{11} \\ \varepsilon_{22} \\ \varepsilon_{33} \\ \varepsilon_{23} \\ \varepsilon_{31} \\ \varepsilon_{12} \end{pmatrix}. \qquad (5.58)$$

The matrix in (5.58) is also that of the quadratic form of the energy density, provided we use the variables $(\varepsilon_{11}, \varepsilon_{22}, \varepsilon_{33}, \gamma_{23}, \gamma_{31}, \gamma_{12})$.

## 5.9. Infinitesimal Stability

We shall now discuss the stability of the initial strain state $\varepsilon_{ij} \equiv 0$ in the possible presence of initial stresses. We define such stability by the requirement that the strain energy

$$U = \int_V (\tau^0_{ij}\varepsilon_{ij} + \frac{1}{2} C^{pq}_{ij}\varepsilon_{ij}\varepsilon_{pq})dV \geq 0 \tag{5.59}$$

have a minimum for $\varepsilon_{ij} \equiv 0$ with respect to all possible infinitesimal changes of configuration for which there exists a univalent displacement field related to the strains by

$$\varepsilon_{ij} = \frac{1}{2}(D_iu_j + D_ju_i). \tag{5.60}$$

The equality to zero cannot occur in (5.59) unless $u_i$ in (5.60) is an infinitesimal rigid body displacement, in which case we know that the strain field vanishes identically. We assume that no kinematic constraint is imposed on the boundary $\partial V$ which would either modify or prohibit the existence of such rigid displacements. Indeed, the boundary will be regarded as entirely free.

Condition (5.59) requires that the linear part vanish and that the quadratic part be positive definite in the variable $\varepsilon_{ij}$ subject to (5.60). The vanishing of the linear part

$$\int_V \tau^0_{ij} \frac{1}{2}(D_iu_j + D_ju_i)dV = \int_V \tau^0_{ij}D_iu_j \, dV = 0$$

for every field $u_j$ with continuous first derivatives implies, after an integration by parts,

$$\int_{\partial V}(n_i\tau^0_{ij})u_j \, dS - \int_V u_j(D_i\tau^0_{ij}) \, dV = 0.$$

This requires that the initial stress field be *self-equilibrated*, i.e., at each interior point the initial stresses must satisfy the translational equilibrium equations without body forces

$$D_i \tau^o_{ij} = 0 , \tag{5.61}$$

and at each boundary point the traction-free conditions

$$n_i \tau^o_{ij} = 0 . \tag{5.62}$$

The initial stresses of a stable initial configuration cannot be *relaxed* (completely annulled) by a change of configuration preserving the physical continuity of the medium. Indeed, for that to be possible we would need to find a field of displacements $u^o_q$ such that the associated strains

$$\varepsilon_{pq} = \varepsilon^o_{pq} = \frac{1}{2} (D_p u^o_q + D_q u^o_p)$$

when used in (5.54) with $\tau_{ij} = 0$, would yield

$$\tau^o_{ij} = -C^{pq}_{ij} \varepsilon^o_{pq} . \tag{5.63}$$

For such a field, however, (5.59) would become

$$U^o = - \frac{1}{2} \int_V C^{pq}_{ij} \varepsilon^o_{ij} \varepsilon^o_{pq} \, dV \geq 0,$$

and this would be inconsistent with the positive definiteness of the quadratic part of the energy, unless it were to vanish. But vanishing would imply that $\varepsilon^o_{pq} \equiv 0$ and that the initial stresses would already vanish at the outset.

One is forced to conclude that a relaxation of the initial stresses implies a dislocation of the continuous medium. If the region is multiply connected, we have already met dislocations of the Weingarten-Volterra type capable of explaining the presence of fields $\varepsilon^o_{pq}$ which, although regular, cannot be relaxed by a change of configuration.

One may imagine, in a much more general way, even in simply connected regions, fields $\varepsilon^{o}_{pq}$ connected by (5.63) with the initial stresses and incapable of relaxation because they do not meet conditions (5.10) for local compatibility. The initial stresses may thus be associated with nonvanishing components of the incompatibility tensor and, by their relaxation, cause dislocations of a much more complex nature than those of Weingarten-Volterra.

## 5.10. Hadamard's Condition for Infinitesimal Stability

In view of the symmetry properties (5.56) the second part of the criterion of infinitesimal stability may be given the form

$$\frac{1}{2}\int_V C^{pq}_{ij}\varepsilon_{ij}\varepsilon_{pq}\,dV = \frac{1}{2}\int_V C^{pq}_{ij}D_i u_j D_p u_q\,dV \geq 0 \qquad (5.64)$$

and is a fundamental constraint imposed on the moduli of elasticity of the medium. This constraint may be used easily in *homogeneous* elastic media, i.e., those characterized by a matrix of moduli which is the same at every point. If we define a *stable elastic medium* as one which satisfies (5.64) for all sufficiently smooth $u$; the following theorem may readily be established.

Theorem. A homogeneous elastic medium with no kinematic boundary conditions is stable if and only if the matrix of moduli is positive definite.

The sufficiency of the condition is obvious from (5.64). It is also necessary because, with no kinematic condition on the boundaries, one may use a linear displacement field

$$u_j = \alpha_j + \alpha_{ji}x_i$$

for which the strain field

$$\varepsilon_{ij} = \frac{1}{2}(\alpha_{ji} + \alpha_{ij})$$

is arbitrary and constant. In this instance the condition for stability reduces to the requirement that

$$\frac{1}{2} C_{ij}^{pq}\varepsilon_{ij}\varepsilon_{pq} \int_V dV$$

be positive definite, and this clearly implies the positive definiteness of $C_{ij}^{pq}$.

When the medium is inhomogeneous, it is a difficult problem to find local conditions on the matrices of the moduli necessary and sufficient for stability. A necessary local condition has been established by Hadamard for the case in which the variation in the matrix from point to point is sufficiently regular. For this purpose he considers a perturbation of displacement localized in a block near the point of interest, which may without loss of generality be chosen as the origin. The displacement has the form $u_j = a_j\phi(x)$ where $a_j$ is an arbitrary constant vector and

$$\phi(x) = \begin{cases} (x_1^2-\varepsilon_1^2)^2(x_2^2-\varepsilon_2^2)^2(x_3^2-\varepsilon_3^2)^2, & -\varepsilon_i \le x_i \le \varepsilon_i \\ 0 \quad \text{otherwise} & . \end{cases}$$

The displacement and its gradients are thus continuous on the boundaries of the block. We write

$$C_{ij}^{pq}a_j a_q = S_i^p$$

which is a symmetric, second order Cartesian tensor; one may therefore always regard the axes as oriented along the principal directions of this tensor at the origin. If $S_i^p$ is sufficiently regular at the origin, then as $\varepsilon_i$ tend to zero the second member of (5.64) approaches the limit

$$\frac{1}{2} S_1\int_V (D_1\phi)^2 dV + \frac{1}{2} S_2\int_V (D_2\phi)^2 dV + \frac{1}{2} S_3\int_V (D_3\phi)^2\ dV.$$

where $S_i$ are the eigenvalues of $S_i^p$. The integrals are elementary and extend only through the block. One finds

$$\int_V (D_i\phi)^2 dV = 3\left(\frac{256}{5\cdot 7\cdot 9}\right)^3 \frac{(\varepsilon_1\ \varepsilon_2\ \varepsilon_3)^9}{\varepsilon_i^2}$$

and concludes that the expression

$$\frac{S_1}{\varepsilon_1^2} + \frac{S_2}{\varepsilon_2^2} + \frac{S_3}{\varepsilon_3^2}$$

should be positive definite, which implies

$$S_1 > 0, \quad S_2 > 0, \quad S_3 > 0.$$

Because its eigenvalues are positive, the tensor $S_i^p$ is thus positive definite; equivalently, the quadratic form $S_i^p b_i b_p$ is positive definite. Finally, then, the necessary condition for Hadamard's infinitesimal stability is that, at every point, the biquadratic form

$$C_{ij}^{pq} b_i b_p a_j a_q \tag{5.65}$$

be positive definite.

Compared with the local requirement that

$$C_{ij}^{pq} \varepsilon_{ij} \varepsilon_{pq} \tag{5.66}$$

be positive definite, which is itself sufficient, Hadamard's condition is weaker. The proof of this assertion begins by using (5.56) to reduce (5.65) to the equivalent requirement that

$$C_{ij}^{pq}(b_i a_j + b_j a_i)(b_p a_q + b_q a_p)$$

be positive definite. In obvious matrix notation the tensor $b_i a_j + a_i b_j$ may be written $T = \mathbf{ba} + \mathbf{ab}$. Its characteristic equation $Tc = \lambda c$ always has a solution $\lambda = 0$ for a principal direction $c$ which is orthogonal simultaneously to $a$ and $b$. Since symmetric tensors $T$ with one vanishing eigenvalue form a proper subset of the set of all symmetric tensors, it follows that the inequalities which are imposed on the moduli of elasticity in order to meet Hadamard's condition are less restrictive than those assuring that their matrix be positive definite.

We have seen that Hadamard's condition is not sufficient for infinitesimal stability in a homogeneous medium. The study of the inhomogeneous cases for which it will suffice remains to be undertaken.

## 5.11. Isotropy and Anisotropy

Equations (5.54), (5.61), and (5.62) allow a slight simplification of linear elasticity: they imply that the determination of initial stresses and of further stresses caused by changes of configuration are separable problems. Because the initial stresses in a stable configuration form a self-equilibrated field, the additional stresses satisfy the same equilibrium equations with the exterior forces as the total stresses. The problem of determining additional stresses is thus not altered by the hypothesis that the initial state is the *natural state*, i.e., that the initial stresses vanish. This amounts to retaining only the quadratic part of the strain energy density, (i.e., that depending on the moduli of elasticity), as we shall henceforth do.

The determination of the initial stresses is an extremely complex problem in the physics of materials, on the theoretical level as well as the experimental. In the rare cases in which the initial stresses are regarded as known, it is enough to add them to the additional stresses to obtain the total stresses.

We will now consider some special hypotheses of material symmetry which simplify the tensor of the moduli of elasticity. The first is *isotropy*, the assumption that the material offers no preferential direction. The strain energy of an isotropic elastic material should be expressible as a function of the fundamental invariants of strain in such a way as to provide the same formulation regardless of the orientation chosen for the Cartesian frame. Since we allow only homogeneous quadratic forms for the strain energy density, we need consider only two distinct combinations of fundamental invariants. By starting with the homogeneous quadratic combinations $I_1^2$ and $\Gamma_2$, one may write the strain energy density in terms of two moduli $K$ and $G$ as

$$W = \frac{1}{2} K I_1^2 - 2G \Gamma_2. \qquad (5.67)$$

We recall that

$$I_1 = \varepsilon_{11} + \varepsilon_{22} + \varepsilon_{33}$$

and note that, from (5.31), one may write

$$\begin{aligned} -2\Gamma_2 &= \tfrac{1}{3}(\varepsilon_{11}-\varepsilon_{22})^2 + \tfrac{1}{3}(\varepsilon_{22}-\varepsilon_{33})^2 + \tfrac{1}{3}(\varepsilon_{33}-\varepsilon_{11})^2 \\ &\quad + \tfrac{1}{2}(\gamma_{23}^2 + \gamma_{31}^2 + \gamma_{12}^2) \end{aligned}$$

so that the application of formulas (5.48) yields the explicit constitutive relations

$$\tau_{11} = (K + \tfrac{4}{3} G)\varepsilon_{11} + (K - \tfrac{2}{3} G)(\varepsilon_{22} + \varepsilon_{33}) \qquad (5.68)$$

for a normal stress and

$$\tau_{23} = G\,\gamma_{23} \qquad (5.69)$$

for a shear stress, the other elements of the tensor being obtained simply by cyclic permutation of the indices.

The last relation justifies calling $G$ the *shear modulus*. On the other hand, by forming the arithmetic mean of the normal stresses

$$\frac{\tau_{11} + \tau_{22} + \tau_{33}}{3} = K(\varepsilon_{11} + \varepsilon_{22} + \varepsilon_{33}) = K\, I_1 \qquad (5.70)$$

and recalling the interpretation of the first invariant as change of specific volume, one reaches the justification for calling $K$ the *bulk modulus*.

An isotropic material can thus be characterized by these two moduli. The condition

$$G > 0, \quad K > 0, \qquad (5.71)$$

is necessary and sufficient for the strain energy density to be locally positive definite. The matrix of the moduli takes the form

$$\begin{pmatrix} K + 4G/3 & K - 2G/3 & K - 2G/3 & 0 & 0 & 0 \\ K - 2G/3 & K + 4G/3 & K - 2G/3 & 0 & 0 & 0 \\ K - 2G/3 & K - 2G/3 & K + 4G/3 & 0 & 0 & 0 \\ 0 & 0 & 0 & G & 0 & 0 \\ 0 & 0 & 0 & 0 & G & 0 \\ 0 & 0 & 0 & 0 & 0 & G \end{pmatrix}. \qquad (5.72)$$

The alternative system of *Lamé's* moduli is connected with the foregoing by

$$\lambda = K - \frac{2}{3} G, \qquad \mu = G \qquad (5.73)$$

and gives the constitutive equations the simple form

$$\tau_{ij} = \lambda \varepsilon_{qq} \delta_{ij} + 2\mu \varepsilon_{ij}. \qquad (5.74)$$

This is equivalent to defining the general moduli by

$$C_{ij}^{pq} = \lambda\delta_{ij}\delta_{pq} + \mu(\delta_{ip}\delta_{jq} + \delta_{jp}\delta_{iq}), \tag{5.75}$$

in close accord with the required symmetry properties and the result that an isotropic fourth order tensor must be a linear combination of the isotropic tensors $\delta_{ij}\delta_{pq}$, $\delta_{ip}\delta_{jq}$, and $\delta_{jp}\delta_{iq}$.

Formula (5.75) gives quick insight into Hadamard's condition for infinitesimal stability (5.65), which reduces here to

$$\lambda(\mathbf{a}\cdot\mathbf{b})^2 + 2\mu(\mathbf{a}\cdot\mathbf{a})(\mathbf{b}\cdot\mathbf{b}) \geq 0$$

or again, if $\theta$ is the angle between the two vectors,

$$\lambda \cos^2\theta + 2\mu > 0.$$

The latter form yields Hadamard's conditions

$$\mu > 0, \quad \lambda + 2\mu > 0.$$

In terms of the bulk and shear moduli, they become

$$G > 0, \quad K > -\tfrac{4}{3}G, \tag{5.77}$$

and these hold if and only if the diagonal terms of the matrix (5.72) are positive. We note that Hadamard's conditions do not rule out a negative bulk modulus, but do bound it below. It may be shown that they coincide exactly with those for the existence of real propagation speeds of compressional and equivoluminal waves in the medium. This observation will be enlarged when we see the effect of Hadamard's general condition in the elliptic character of the partial differential equations of linear elasticity.

The system of moduli $(G,K)$ is conceptually simplest

for an isotropic material, and gives meaning to the decomposition of states of strain and stress into an isotropic (or spherical) part and a deviator, for one sees easily from (5.68) and (5.69) that the corresponding invariants are pairwise proportional:

$$J_1 = 3K\, I_1 \qquad \Sigma_2 = 4G^2\, \Gamma_2. \tag{5.78}$$

Thus in an isotropic medium the mechanism of distortionless volume change under hydrostatic stress and that of equivoluminal distortion under the action of the stress deviator operate independently. The principal directions of strain coincide locally everywhere with those of stress. This is not generally true in anisotropic media.

In technical applications one rarely uses the system of moduli $(G,K)$ or that of Lamé. This stems from the experimental determination of the moduli by applying axial tractions to the ends of a prismatic bar. If the $x_1$ axis lies along the axis of the bar, then the only non-vanishing components of stress is $\tau_{11} = F/S$, where $F$ is the resultant of the axial tractions and $S$ is the cross-sectional area of the bar.

Therefore, by formulas (5.69) $\gamma_{23} = \gamma_{31} = \gamma_{12} = 0$, and by formulas (5.68)

$$\tau_{11} = (K + \tfrac{4}{3}G)\varepsilon_{11} + (K - \tfrac{2}{3}G)(\varepsilon_{22} + \varepsilon_{33})$$

$$0 = (K + \tfrac{4}{3}G)\varepsilon_{22} + (K - \tfrac{2}{3}G)(\varepsilon_{33} + \varepsilon_{11})$$

$$0 = (K + \tfrac{4}{3}G)\varepsilon_{33} + (K - \tfrac{2}{3}G)(\varepsilon_{11} + \varepsilon_{22}).$$

By subtracting the last two from each other we get

$$\varepsilon_{22} = \varepsilon_{33}$$

and are led to define a coefficient $\nu$ by

$$\varepsilon_{22} = \varepsilon_{33} = -\nu\varepsilon_{11} \quad \text{with} \quad \nu = \frac{K - \frac{2}{3}G}{2(K + \frac{1}{3}G)} . \tag{5.79}$$

This parameter is called *Poisson's ratio*; it measures the reduction in transversal dimensions connected experimentally with elongation of the bar. It is positive for all known bodies, indicating that normally $3K > 2G$. Its limit as $K$ tends to infinity is $1/2$; this is Poisson's ratio for incompressible media. The formula for $\tau_{11}$ can now be written

$$\tau_{11} = E\varepsilon_{11} \quad \text{with} \quad E = \frac{9KG}{3K+G} = 2(1+\nu)G. \tag{5.80}$$

The parameter $E$ is called the *modulus of elasticity* or *Young's modulus*.

Inversion of relations (5.79) and (5.80) yields

$$K = \frac{E}{3(1-2\nu)}, \qquad G = \frac{E}{2(1+\nu)} , \tag{5.81}$$

allowing expression of the constitutive equations (5.68) and (5.69) in terms of Young's modulus and Poisson's ratio. These equations become singular in the limiting case of incompressibility, where $\nu = 1/2$. On the other hand, solved for the strains and known as *Hooke's law*, the equations

$$\begin{aligned} \varepsilon_{11} &= \frac{1}{E}[\tau_{11} - \nu(\tau_{22} + \tau_{33})], \quad \varepsilon_{22} \; \ldots\ldots , \\ \gamma_{23} &= \frac{2(1+\nu)}{E}\tau_{23}, \qquad \gamma_{31} = \ldots , \end{aligned} \tag{5.82}$$

are not singular.

In an anisotropic medium the expression for the strain energy density cannot be completely independent of the orientation of the frame. If the properties of the medium are invariant with respect to certain rotations of the frame, there is reason to seek the combinations of the strain tensor invariant under those rotations. The search may be conducted by a process generalizing that which leads to the fundamental invariants.

Let $\mathsf{E}$ be the matrix of the strain tensor. Consider the response of $\det(\mathsf{E}-\mathsf{Q})$ to a change in the orientation of the frame produced by a rotation with matrix $\mathsf{R}$. Using the matrix form of (4.15), we find

$$\det(\mathsf{E}-\mathsf{Q}) = \det(\mathsf{R}^T\hat{\mathsf{E}}\mathsf{R}-\mathsf{Q}) = \det[\mathsf{R}^T(\hat{\mathsf{E}} - \mathsf{R}\mathsf{Q}\mathsf{R}^T)\mathsf{R}]$$
$$= \det(\hat{\mathsf{E}}-\mathsf{R}\mathsf{Q}\mathsf{R}^T).$$

It follows that if the matrix $\mathsf{Q}$ is invariant (i.e., if $\mathsf{R}\mathsf{Q}\mathsf{R}^T = \mathsf{Q}$) then the expansion of the determinant will produce a polynomial in the elements of $\mathsf{Q}$, having coefficients of combinations of the elements of $\mathsf{E}$ invariant under that rotation.

We apply this notion first to the case of *transverse isotropy*, which obtains when the medium has a privileged direction such that all directions orthogonal to it are equivalent with regard to elastic properties. With this first axis of the frame along the priviledged direction, elastic properties will remain invariant under rotations of the form

$$\mathsf{R} = \begin{pmatrix} 1 & 0 & 0 \\ 0 & \cos\alpha & \sin\alpha \\ 0 & -\sin\alpha & \cos\alpha \end{pmatrix}$$

where $\alpha$ is arbitrary. The most general matrix $\mathsf{Q}$ invariant under this group of rotations is

$$\mathsf{Q} = \begin{pmatrix} a & 0 & 0 \\ 0 & b & c \\ 0 & -c & b \end{pmatrix},$$

where a, b, and c are arbitrary. Hence the expression

$$\begin{aligned}\det(E-\mathfrak{a}) &= a(b^2+c^2) + ab(\varepsilon_{22} + \varepsilon_{33}) + c^2\varepsilon_{11} \\ &\quad + a(\varepsilon_{23}^2 - \varepsilon_{22}\varepsilon_{33}) + b(\varepsilon_{12}^2+\varepsilon_{13}^2-\varepsilon_{11}(\varepsilon_{22}+\varepsilon_{33})) \\ &\quad + \det E\end{aligned}$$

is invariant, whatever the parameters $a$, $b$, $c$ may be.

In this way one finds two invariants of the first degree, the coefficients of $ab$ and $c^2$ (whose sum yields the first fundamental invariant); two invariants of the second degree, the coefficients of $a$ and $b$ (whose sum yields the second fundamental invariant); and the invariant $I_3$ as an independent term. The quadratic form for the strain energy density must depend on the strain only through these invariants. Thus it must take the form

$$\begin{aligned}W = \tfrac{1}{2}A\varepsilon_{11}^2 + \tfrac{1}{2}B(\varepsilon_{22} + \varepsilon_{33})^2 + C\varepsilon_{11}(\varepsilon_{22} + \varepsilon_{33}) \\ + 2D(\varepsilon_{23}^2 - \varepsilon_{22}\varepsilon_{33}) + 2E(\varepsilon_{12}^2 + \varepsilon_{13}^2)\end{aligned}$$

and this generates a matrix of moduli

$$\begin{pmatrix} A & C & C & 0 & 0 & 0 \\ C & B & B-2D & 0 & 0 & 0 \\ C & B-2D & B & 0 & 0 & 0 \\ 0 & 0 & 0 & D & 0 & 0 \\ 0 & 0 & 0 & 0 & E & 0 \\ 0 & 0 & 0 & 0 & 0 & E \end{pmatrix}.$$

It is positive definite if all the principal determinants one can form from it are positive. For this it is sufficient to ensure the positive character of a nested sequence of principal determinants. Here, for example, one obtains the conditions

$$A > 0, \quad B > D > 0, \quad E > 0, \quad A(B-D) > 4C^2.$$

Elastic properties of *orthotropic* materials are symmetric with respect to three orthogonal planes, and the axes are chosen as the intersections of these planes. A rotation of $180^\circ$ about the third axis is permitted, and the corresponding rotation

$$\mathsf{R} = \begin{pmatrix} -1 & 0 & 0 \\ 0 & -1 & 0 \\ 0 & 0 & 1 \end{pmatrix}$$

leaves invariant matrices of the type

$$\mathfrak{a} = \begin{pmatrix} a & d & 0 \\ e & b & 0 \\ 0 & 0 & c \end{pmatrix}$$

A rotation of $180^\circ$ about the first axis is also permitted, so the preceding invariant matrices are reduced to those for which $d = e = 0$, and one sees that they are then also invariant under a rotation of $180^\circ$ about the second axis.

The general invariant expression here is

$$\begin{aligned} \det(\mathsf{E}-\mathfrak{a}) &= -abc + bc\varepsilon_{11} + ca\varepsilon_{22} + ab\varepsilon_{33} \\ &\quad + a(\varepsilon_{23}^2 - \varepsilon_{22}\varepsilon_{33}) + b(\varepsilon_{31}^2 - \varepsilon_{33}\varepsilon_{11}) + c(\varepsilon_{12}^2 - \varepsilon_{11}\varepsilon_{22}) \\ &\quad + \det \mathsf{E}. \end{aligned}$$

Here we have three invariants of the first degree and three of the second, implying for the strain energy density a quadratic form with nine moduli

$$\begin{aligned} W = {} & \tfrac{1}{2}(A_{11}\varepsilon_{11}^2 + A_{22}\varepsilon_{22}^2 + A_{33}\varepsilon_{33}^2 + 2A_{23}\varepsilon_{22}\varepsilon_{33} + 2A_{31}\varepsilon_{33}\varepsilon_{11} + 2A_{12}\varepsilon_{11}\varepsilon_{22}) \\ & + 2G_{23}(\varepsilon_{23}^2 - \varepsilon_{22}\varepsilon_{33}) + 2G_{31}(\varepsilon_{31}^2 - \varepsilon_{33}\varepsilon_{11}) + 2G_{12}(\varepsilon_{12}^2 - \varepsilon_{11}\varepsilon_{22}), \end{aligned}$$

with the matrix

$$\begin{bmatrix} A_{11} & A_{12}-2G_{12} & A_{13}-2G_{13} & 0 & 0 & 0 \\ A_{21}-2G_{21} & A_{22} & A_{23}-2G_{23} & 0 & 0 & 0 \\ A_{31}-2G_{31} & A_{32}-2G_{32} & A_{23} & 0 & 0 & 0 \\ 0 & 0 & 0 & G_{23} & 0 & 0 \\ 0 & 0 & 0 & 0 & G_{31} & 0 \\ 0 & 0 & 0 & 0 & 0 & G_{12} \end{bmatrix}$$

which is positive definite under the conditions

$$G_{23} > 0, \quad G_{31} > 0, \quad G_{12} > 0, \quad A_{11} > 0, \quad A_{11}A_{22} > (A_{12}-2G_{12})^2,$$

$$A_{11}A_{22}A_{33} + (A_{23}-2G_{23})(A_{31}-2G_{31})(A_{12}-2G_{12}) > A_{11}(A_{23}-2G_{23})^2$$
$$+ A_{22}(A_{31}-2G_{31})^2 + A_{33}(A_{12}-2G_{12})^2.$$

If the first axis coincides with an axis of *ternary symmetry* (i.e., symmetry with respect to rotations of $120^\circ$ about the axis), then one sees easily that the only matrices $\mathfrak{a}$ invariant under the operator

$$R = \begin{pmatrix} 1 & 0 & 0 \\ 0 & \cos 120^\circ & \sin 120^\circ \\ 0 & -\sin 120^\circ & \cos 120^\circ \end{pmatrix} = -\frac{1}{2}\begin{pmatrix} -2 & 0 & 0 \\ 0 & 1 & -\sqrt{3} \\ 0 & \sqrt{3} & 1 \end{pmatrix}$$

are those invariant under rotation through an arbitrary angle about the same axis. Ternary symmetry is equivalent, therefore, to transverse isotropy with respect to that axis.

## 5.12. Criteria for Elastic Limits

For most materials, especially metals, behavior may be regarded as elastic in a region of the stress state which is sufficiently restricted, and it is important to know these limits. Here we develop only a few classical criteria related to *isotropic* materials.

The *Tresca criterion* (1864), reexamined by *Guest* (1900), asserts that the limit is defined by a certain value of the maximum shear stress. In a simple tension test the state of stress is uniaxial; the axis of the bar is principal and corresponds to the stress $\sigma_1 = F/S$, where $F$ is the force and $S$ the cross-sectional area, while $\sigma_2 = \sigma_3 = 0$. The theory of Mohr's circles then indicates the value $\sigma_1/2$ for the maximum shear stress. If the analysis of the tension test leads to limiting the elastic region to $|\sigma_1| < \sigma_e$, a value known as the elastic limit, the Tresca criterion applied to a *general* stress state requires that at each point

$$\sigma_1 - \sigma_3 < \sigma_e, \tag{5.83}$$

where $\sigma_1$ and $\sigma_3$ are the maximum and minimum principal stress, respectively.

Indeed, the isotropy of the medium requires that the criterion be expressed in terms of invariants of the stress state. Application of the Tresca criterion (5.83) at a point requires a knowledge of the principal stresses. This usually necessitates solving a third degree equation.

The most general form for a criterion for isotropic material would be

$$f(J_1, J_2, J_3) < 0.$$

Using the fact that, within certain limits, experience indicates that the elastic behavior of a metal is unaffected by superposing a state of *hydrostatic* (spherical) stress, *von Mises* (1913) suggested using only the invariants of the deviator. In fact, he proposed the criterion

$$\begin{aligned} -6\Sigma_2 &= (\tau_{11}-\tau_{22})^2 + (\tau_{22}-\tau_{33})^2 + (\tau_{33}-\tau_{11})^2 + 6(\tau_{23}^2+\tau_{21}^2+\tau_{12}^2) \\ &= (\sigma_1-\sigma_2)^2 + (\sigma_2-\sigma_3)^2 + (\sigma_3-\sigma_1)^2 < 2\sigma_e^2. \end{aligned} \tag{5.84}$$

*Hencky* (1924) interprets this criterion as restricting, in (5.67), the "distortion" part of the strain energy density

$$-2G\Gamma_2 = -\frac{1}{2G}\Sigma_2 < \frac{1}{6G}\sigma_e^2.$$

*Nadai* (1937) interprets it using the idea of octahedral shear stress defined by (5.39); then it takes the form

$$|t_{ot}| < \frac{\sqrt{2}}{3}\sigma_e. \tag{5.85}$$

The Hüber-Hencky-von Mises criterion is easy to apply because it is expressed rationally in terms of the elements of the stress state for any frame whatever.

For most metals it is in better agreement with experiment than Tresca's criterion.

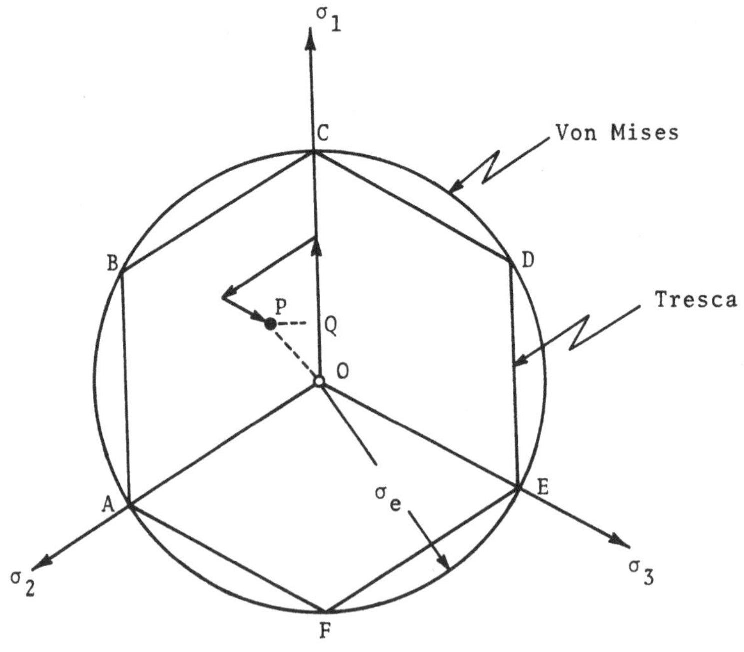


Figure 5.5

The difference between the two criteria may be illustrated simply by a graphical construction (Figure 5.5). Consider a system of three coplanar unit vectors $\mathbf{g}_i$, making angles of $120^{\circ}$ with each other, so that

$$\mathbf{g}_i \cdot \mathbf{g}_j = \begin{cases} 1 & \text{if } i = j \\ \cos 120^{\circ} = -\frac{1}{2} & \text{if } i \neq j \end{cases} . \tag{5.86}$$

The point representing a stress state with principal values $(\sigma_1, \sigma_2, \sigma_3)$ is defined by the position vector

$$\mathbf{s} = \sigma_1\mathbf{g}_1 + \sigma_2\mathbf{g}_2 + \sigma_3\mathbf{g}_3. \tag{5.87}$$

Stress states differing only by a hydrostatic state have the same representative point because the basis vectors are connected by the relation

$$\mathbf{g}_1 + \mathbf{g}_2 + \mathbf{g}_3 = 0. \tag{5.88}$$

This is unimportant for the two criteria under study, which depend only on the differences between principal stresses. From (5.86) and (5.87) one finds that

$$\begin{aligned} \mathbf{s}\cdot\mathbf{s} &= \sigma_1^2 + \sigma_2^2 + \sigma_3^2 - \sigma_1\sigma_2 - \sigma_2\sigma_3 - \sigma_3\sigma_1 \\ &= \tfrac{1}{2}[(\sigma_1-\sigma_2)^2 + (\sigma_2-\sigma_3)^2 + (\sigma_3-\sigma_1)^2] \end{aligned}$$

and then, in view of (5.84), the von Mises criterion becomes

$$\mathbf{s}\cdot\mathbf{s} = -3\Sigma_2 < \sigma_e^2,$$

so that the region where the elastic limit is not exceeded is a circle of radius $\sigma_e$.

The Pythagorsan theorem allows the calculation of the component $\overline{PQ}$, normal to the $\sigma_1$-axis, of a typical stress point $\mathbf{s}$:

$$\overline{PQ}^2 = \mathbf{s}\cdot\mathbf{s} - (\mathbf{s}\cdot\mathbf{g}_1)^2 = 3\left(\frac{\sigma_2-\sigma_3}{2}\right)^2.$$

In order that the shear stress $\frac{1}{2}(\sigma_2-\sigma_3)$ not exceed Tresca's limit it is necessary that

$$\overline{PQ} < \frac{\sqrt{3}}{2}\,\sigma_e,$$

restricting the elastic region to the strip between the lines AB and ED. Since $\frac{1}{2}(\sigma_3-\sigma_1)$ and $\frac{1}{2}(\sigma_1-\sigma_2)$ may also be the maximum shear stress, the elastic region for Tresca's criterion is the hexagon ABCDEF inscribed in the circle for von Mises' criterion.

Tresca's criterion is clearly slightly more demanding than von Mises', both having been formulated so as to coincide in a uniaxial test like that represented by the point C. We observe that they coincide also in biaxial states ($\sigma_2 = 0$, $\sigma_3 = \sigma_1 = \sigma_e$) like those represented by the point D.

While it is true that materials can endure an apparently unlimited hydrostatic *compression* without suffering permanent deformations, there should be an approximate limit for their collapse in a state of hydrostatic *tension*. Experimental realization of such a state is so difficult that, although knowledge of this limit is important on the theoretical level, it is doubtless less so on the practical level.

One may believe that the state of hydrostatic stress would be taken into consideration by a criterion of the type $\Sigma_2 = f(J_1)$, equivalent in Nadai's interpretation to a relation

$$t_{ot} = f(t_{on}) \tag{5.89}$$

between the octahedral normal and shear stresses. General criteria of this type amount to neglecting the effect of the third fundamental invariant of stress.

Another criterion, due to Mohr, is called that of the *intrinsic curve*. It assumes that the elastic limit is reached

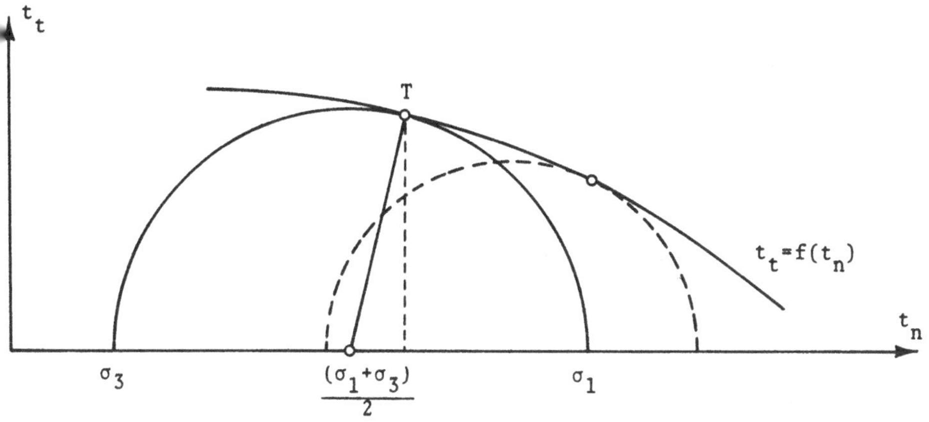


Figure 5.6

on a face when the magnitude of the shear tractions $t_t$ on it reaches a level depending on the normal traction $t_n$. The equation $t_t = f(t_n)$ of the limit is the intrinsic curve of the isotropic material. In Mohr's diagram (Figure 5.6) the curve is an envelope of Mohr's circles for $n_2 = 0$, when the principal stresses are arranged according to the convention (5.43). We thus have the equation

$$t_t^2 + (t_n - \frac{\sigma_1+\sigma_3}{2})^2 = (\frac{\sigma_1-\sigma_3}{2})^2$$

stating that the point of contact on the envelope belongs to Mohr's circle for $n_2 = 0$, and the equation

$$\frac{1}{t_t}(t_n - \frac{\sigma_1+\sigma_3}{2}) = -f'(t_n)$$

stating that the line joining the center of the circle to the point of contact is perpendicular to the tangent to the intrinsic curve. With $t_t$ replaced by $f(t_n)$ this equation becomes

$$\frac{\sigma_1+\sigma_3}{2} = t_n + f\,f' \tag{5.90}$$

and the prceeding one becomes

$$\frac{\sigma_1-\sigma_3}{2} = f\sqrt{1 + f'^2}. \tag{5.91}$$

Equations (5.90) and (5.91) may be regarded as equations in a parameter $t_n$ of a limiting locus

$$\frac{\sigma_1-\sigma_3}{2} = F(\frac{\sigma_1+\sigma_3}{2}) \tag{5.92}$$

which evidently generalizes Tresca's criterion.

## 5.13. Navier's Equations

Navier's equations of linear elasticity are the partial differential equations which control the displacement vector. They are obtained from the translational equilibrium equations by first writing the stresses in terms of the strains by the constitutive equations, and then the strains in terms of the displacement gradients:

$$D_i(C^{pq}_{ij}D_p u_q) + \rho_o g_j = 0. \tag{5.93}$$

These equations are accompanied by boundary conditions which consist in imposing some displacements

$$u_q = \bar{u}_q \tag{5.94}$$

on a part $\partial_1 V$ of the boundary, and some surface tractions

$$n_i\tau_{ij} = n_i C^{pq}_{ij} D_p u_q = \bar{t}_j \tag{5.95}$$

on the complementary part $\partial_2 V$.

We now establish the *elliptic* character of these partial differential equations. According to the method of *Cauchy-Kowalewska,* it suffices to show that the *simultaneous* assignment of the values $u_q$ and $t_j$ on a part of the boundary always allows the calculation of the partial derivatives of the displacement to any desired order. On that portion of the surface the displacements will have a complete Taylor expansion and, if it converges, the solution of Navier's equations will be analytic in a neighborhood of that portion of the surface.

A few special notations will be helpful in the proof. If $n_i$ denotes, as usual, the direction cosines of the normal to the surface, we introduce

$$\Delta = n_i D_i \tag{5.96}$$

for the derivative along that normal. Three derivatives in a plane tangent to the surface can be formed by the operators

$$\partial_s = e_{smr} n_m D_r. \tag{5.97}$$

They are clearly not independent, as one sees from the identity

$$n_s \partial_s \equiv 0. \tag{5.98}$$

Each of these derivatives is taken in a direction perpendicular to both the normal and one of the Cartesian axes. A partial derivative along a Cartesian axis may now be expressed as

$$D_p = n_p \Delta - e_{pts} n_t \partial_s, \tag{5.99}$$

as one readily verifies by substituting from (5.96) and (5.97) and recalling that $n_t n_t = 1$.

The values of $u_q$ being given on the surface, the derivatives $\partial_s u_q$ in the tangent plane are known. Using (5.99) in the expressions (5.95) for $t_j$ on the surface, and bringing known terms to the right-hand side, yields

$$n_i n_p C_{ij}^{pq} \Delta u_q = t_j + e_{pts} n_t n_i C_{ij}^{pq} \partial_s u_q, \tag{5.100}$$

a system of three equations for the three still unknown first partial derivatives of the $u_q$. According to Hadamard's condition of infinitesimal stability, the matrix

$$n_i n_p C_{ij}^{pq} = S_j^q$$

of this system is positive definite and therefore invertible. The three remaining derivatives are thus uniquely determined. Navier's equations now yield

$$C_{ij}^{pq} D_p D_i u_q = -\rho_0 g_j - D_p u_q D_i C_{ij}^{pq}$$

where the right-hand side is known. Once again, the derivatives $D_i u_q$ being known on the surface, the same is true of the second derivatives $\partial_s D_i u_q$, and substituting $D_p$ yields

$$n_p C_{ij}^{pq} \Delta D_i u_q = e_{pts} n_t C_{ij}^{pq} \partial_s D_i u_q - \rho_0 g_j - D_p u_q D_i C_{ij}^{pq}$$

with a known right hand side. Calculation of the still unknown derivatives $\Delta D_i u_q$ comes down finally to that of $\Delta\Delta u_q$ as we see from the equations

$$\begin{aligned} \Delta D_i u_q = n_r D_r D_i u_q &= D_i(n_r D_r u_q) - D_i n_r \cdot D_r u_q \\ &= D_i \Delta u_q - D_i n_r \cdot D_r u_q \\ &= (n_i \Delta - e_{its} n_t \partial_s) \Delta u_q - D_i n_r \cdot D_r u_q \\ &= n_i \Delta\Delta u_q - (e_{its} n_t \partial_s \Delta u_q + D_i n_r \cdot D_r u_q), \end{aligned}$$

where only the first term of the right hand side is unknown. Consequently, substitution from this equation into Navier's yields a result of the type

$$n_i n_p C_{ij}^{pq} \Delta\Delta u_q = a_j.$$

These equations, with known right hand side, are of the

same type as (5.100) and always supply a unique solution for the still unknown second derivatives. One has thus succeeded in calculating on the surface the set of all second partial derivatives of $u_q$. The process may be repeated indefinitely by taking successive partial derivatives of Navier's equations. It is obviously necessary for this that the field of moduli of elasticity and the curvature tensor of the surface should have derivatives of all orders on the surface.

Thus we have shown that Hadamard's necessary condition for infinitesimal stability is also a sufficient condition for the ellipticity of Navier's equations.

On the other hand, as we shall now demonstrate, the *uniqueness* of the solution of Navier's equations under their boundary conditions follows upon verification of the criterion of infinitesimal stability, and is therefore assured by every system of conditions sufficient for that purpose; in particular, by the local positive definite character of the energy density. Indeed, since the equations are linear, uniqueness is assured if the homogeneous system

$$D_i(C_{ij}^{pq} D_p u_q) = 0,$$

$$u_j = 0 \qquad \text{on} \quad \partial_1 V,$$

$$n_i C_{ij}^{pq} D_p u_q = 0 \qquad \text{on} \quad \partial_2 V,$$

has only the trivial solution. By multiplying each homogeneous Navier equation by the corresponding component of displacement and integrating, we get

$$\int_V u_j D_i(C_{ij}^{pq} D_p u_q)\, dV = 0.$$

After integration by parts, this becomes

$$\int_{\partial V} u_j(n_i C_{ij}^{pq} D_p u_q)\, dS - \int_V C_{ij}^{pq} D_i u_j D_p u_q\, dV = 0.$$

The first integral vanishes because of the homogeneous boundary conditions. If the condition for infinitesimal stability is satisfied, the second can vanish only if

$$\varepsilon_{ij} = \frac{1}{2}(D_i u_j + D_j u_i) \equiv 0.$$

By a mild extension of the usual interpretation, we consider this result as corresponding to the trivial solution: On one hand the homogeneous kinematic conditions imposed on $\partial_1 V$ may be sufficient to exclude the existence of every kind of rigid displacement (as is the case in most problems) and $\varepsilon_{ij} \equiv 0$ implies the solution $u_j \equiv 0$, properly called trivial; otherwise these conditions are insufficient or inapplicable and the solution of the initial problem is defined only up to a rigid type of displacement.

In the special case of a homogeneous isotropic medium, where the moduli are constant and are given by equations (5.75), Navier's equations without body forces take the simple form

$$(\lambda + \mu) D_j (D_i u_i) + \mu D_i D_i u_j = 0, \tag{5.101}$$

where

$$D_i D_i = \frac{\partial^2}{\partial x_1^2} + \frac{\partial^2}{\partial x_2^2} + \frac{\partial^2}{\partial x_3^2} = \nabla^2$$

is the harmonic operator. By multiplying each equation by $D_j$ and adding, then suppressing the factor $\lambda + 2\mu > 0$, we get

$$\nabla^2 I_1 = 0. \tag{5.102}$$

The first invariant is thus a harmonic function. Applying the Laplacian to Navier's equations shows that each component of the displacement satisfies the biharmonic equation

$$\nabla^2\nabla^2 u_j = 0. \tag{5.103}$$

These results illustrate the important role of harmonic and biharmonic operators in all those problems of elasticity in which the medium is taken to be isotropic and homogeneous. They enter as well in the case of body forces which depend on a harmonic potential, as with the gravitational field with no attracting masses:

$$g_j = -D_j\phi \quad \text{with} \quad \nabla^2\phi = 0.$$

## 5.14. The Beltrami-Michell Equations

One may in principle transform the equations of local strain compatibility by substituting in them the stresses in terms of strains obtained by inversion of the constitutive equations. This has been done by *Beltrami* and *Michell* in the case of an isotropic homogeneous medium for which one already has the inverted equations (5.82), which may be combined in the form

$$E\varepsilon_{ij} = (1+\nu)\tau_{ij} - \nu\tau_{rr}\delta_{ij}, \qquad \tau_{rr} = J_1. \tag{5.104}$$

In order to simplify the result by using the equilibrium equations, it is appropriate to modify first the presentation of the compatibility equations. Multiply them by $e_{stq}e_{num}$, use formula (4.18), then put $n = s$ to obtain

$$D_iD_i\varepsilon_{tu} + D_tD_u\varepsilon_{ii} - D_tD_i\varepsilon_{iu} - D_uD_i\varepsilon_{it} = 0.$$

By using (5.104), this becomes

$$(1+\nu)D_iD_i\tau_{tu} - \nu\delta_{tu}D_iD_i\tau_{kk} + D_tD_u\tau_{ii}$$

$$- (1+\nu)D_tD_i\tau_{iu} - (1+\nu)D_uD_i\tau_{it} = 0.$$

The last two terms may be transformed by the equilibrium equations to yield

$$(1+\nu)D_iD_i\tau_{tu} - \nu\delta_{tu}D_iD_i\tau_{kk} + D_tD_u\tau_{ii} + (1+\nu)\rho_o(D_tg_u+D_ug_t) = 0.$$

For $t = u$ one gets an equation permitting elimination of the second term. After division by $(1+\nu)$ the final result is

$$\nabla^2\tau_{tu} + \frac{1}{1+\nu}D_tD_u\tau_{ii} + \rho_o[D_tg_u+D_ug_t + \frac{\nu}{1-\nu}\delta_{tu}D_ig_i] = 0. \quad (5.105)$$

The stress should satisfy these six partial differential equations of the second order along with the three equilibrium equations of the first order. It follows that, for body forces depending on a harmonic potential, the first invariant $J_1$ is a harmonic function and each component of the stress is biharmonic.

It must be kept in mind that if the domain is multiply connected, the conditions of global compatibility also transform into supplementary conditions on the stresses.

# Chapter 6
# Extension, Bending, and Torsion of Prismatic Beams

A prismatic beam is a region bounded by an exterior cylindrical surface, two terminal plane sections perpendicular to the generators and, if the beam is hollow, one or more interior cylindrical surfaces. The axis $Oz$ of an orthogonal Cartesian frame is taken parallel to the generators of the cylinder. In a plane section perpendicular to the generators and referred to axes $Ox$ and $Oy$, the region $D$ is bounded by the exterior directrix $c_o$ and the interior directrices $c_i$ $(i = 1,\dots,n)$ of the cavities (Figure 6.1).

The use of cuts, forming bridges between $c_o$ and each cavity directrix, permits the definition of a complete path, designated by $c$, traversed in the mathematically positive sense along $c_o$, and in the negative sense along the other $c_i$. It is such that the exterior normal $\mathbf{n}$ of the region, the tangent $\mathbf{t}$ in the direction of the path, and the unit vector $\mathbf{e}_z$ form a trihedron with the same orientation as the Cartesian frame.

If $\theta$ denotes the angle between the exterior normal and the axis $Ox$, with the agreed direction of passage, then

$$\cos\theta = \partial x/\partial n = \partial y/\partial s, \quad \sin\theta = \partial y/\partial n = -\partial x/\partial s. \tag{6.1}$$

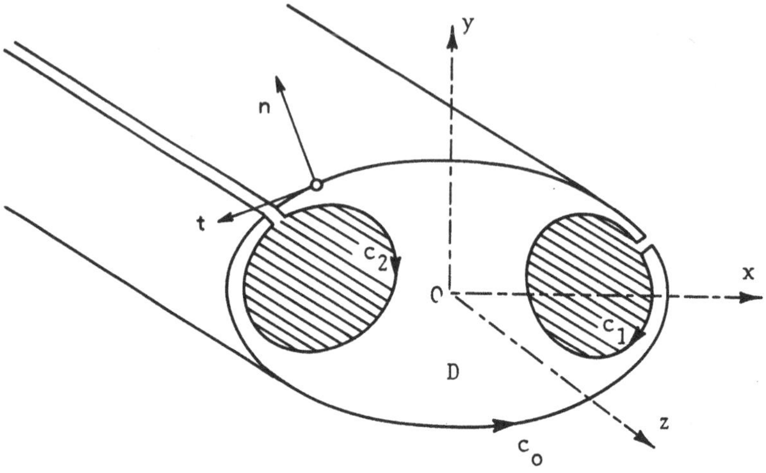


Figure 6.1

## 6.1. Green's and Stokes' Formulas

The theory for isotropic materials as presented in the general equations of Navier and Beltrami-Michell introduces a number of harmonic functions of the variables $(x,y)$. The corresponding versions of Green's and Stokes' formulas are therefore very useful. Let $\mathbf{a}(x,y)$ be a univalent differentiable field of vectors in D. The divergence theorem yields

$$\int_D \operatorname{div} \mathbf{a}\, dx\, dy = \int_C (\mathbf{a}\cdot\mathbf{n})ds = \sum_0^n \int_{C_i} (\mathbf{a}\cdot\mathbf{n})ds$$

because the integrals along the sides of each bridge cancel each other. With $\mathbf{a} = \phi(x,y) \operatorname{grad} \psi(x,y)$ we get Green's first formula

$$\int_D (\phi\nabla^2\psi + \operatorname{grad}\phi \cdot \operatorname{grad}\psi)dxdy = \sum_0^n \int_{C_i} \phi \frac{\partial\psi}{\partial n}\, ds \tag{6.2}$$

where

$$\nabla^2 = \text{div grad} = \partial^2/\partial x^2 + \partial^2/\partial y^2.$$

Green's second formula is found by subtracting from the first the result of interchanging $\phi$ and $\psi$:

$$\int_D (\phi\nabla^2\psi - \psi\nabla^2\phi)\,dxdy = \sum_0^n \int_{c_i} (\phi \frac{\partial\psi}{\partial n} - \psi \frac{\partial\phi}{\partial n})\,ds. \tag{6.3}$$

In both of these formulas we may reverse the direction of the directrices $c_i$ without changing the sign of the right hand side, as long as the normal derivatives are still formed toward the exterior of the region.

Stokes' formula

$$\int_D \mathbf{e}_z \cdot \text{rot}\,\mathbf{a}\,dxdy = \int_C (\mathbf{a}\cdot\mathbf{t})\,ds = \sum_0^n \int_{c_i} (\mathbf{a}\cdot\mathbf{t})\,ds$$

applied also to $\mathbf{a} = \phi$ grad $\psi$ and $\mathbf{a} = \psi$ grad $\phi$ yields

$$\int_D (\frac{\partial\phi}{\partial x}\frac{\partial\psi}{\partial y} - \frac{\partial\phi}{\partial y}\frac{\partial\psi}{\partial x})\,dxdy = \sum_0^n \int_{c_i} \phi\, d\psi = -\sum_0^n \int_{c_i} \psi\, d\phi \tag{6.4}$$

but here it must be noticed that $\partial\psi/\partial s$ and $\partial\phi/\partial s$ change sign with a change in the direction of traversing the path.

## 6.2. The Centroid

In an arbitrarily chosen Cartesian coordinate system, one may find the area $\Omega$ of the domain $D$ and the coordinates $(\hat{x},\hat{y})$ of its *centroid* as follows. With $\phi = x$ and $\psi = y$ in Stokes' formula, one has

$$\Omega = \int_D dx\,dy = \sum_0^n \int_{c_i} x\,dy = -\sum_0^n \int_{c_i} y\,dx. \tag{6.5}$$

Also, with $\phi = \frac{1}{2}x^2$ and $\psi = y$, and with $\phi = x$ and $\psi = \frac{1}{2}y^2$, one has

$$\Omega\,\hat{x} = \int_D x\,dx\,dy = \sum_0^n \int_{c_i} \frac{1}{2}x^2\,dy = -\sum_0^n \int_{c_i} y\,x\,dx,$$
$$\Omega\,\hat{y} = \int_D y\,dx\,dy = \sum_0^n \int_{c_i} xy\,dy = -\sum_0^n \int_{c_i} \frac{1}{2}y^2\,dx. \tag{6.6}$$

In these formulas the directrices are traversed in the direction determined by the complete contour $c$, positive for $c_0$ and reversed for the $c_i$ of the cavities.

We will also need to find the areas and centroids of regions bounded by each one of the separate directrices. If we traverse the boundaries in the mathematically positive sense this yields

$$\Omega_i = \int_{c_i} x\,dy = -\int_{c_i} y\,dx, \qquad i = 0,1,\ldots,n; \tag{6.5'}$$

$$\Omega_i\,\hat{x}_i = \int_{c_i} \frac{1}{2}x^2\,dy = -\int_{c_i} y\,x\,dx,$$
$$\Omega_i\,\hat{y}_i = \int_{c_i} xy\,dy = -\int_{c_i} \frac{1}{2}y^2\,dx, \quad i = 0,1,\ldots,n. \tag{6.6'}$$

Translating the origin of the frame to the centroid of the domain $D$ is an elementary operation. Since this simplifies the statement of results we assume it done and use the corresponding properties

$$\int_D x\,dx\,dy = 0, \qquad \int_D y\,dx\,dy = 0\,. \tag{6.7}$$

## 6.3. Moments of Inertia

The *moments of inertia* of a cross section of the beam with respect to the axes $Ox$ and $Oy$ are defined as

$$I_{xx} = \int_D x^2\, dx\, dy = \frac{1}{3}\sum_0^n \int_{c_i} x^3\, dy = -\sum_0^n \int_{c_i} x^2 y\, dx,$$

$$I_{xy} = \int_D xy\, dx\, dy = -\frac{1}{2}\sum_0^n \int_{c_i} xy^2\, dx = \frac{1}{2}\sum_0^n \int_{c_i} x^2 y\, dy, \tag{6.8}$$

$$I_{yy} = \int_D y^2\, dx\, dy = -\frac{1}{3}\sum_0^n \int_{c_i} y^3\, dx = \sum_0^n \int_{c_i} xy^2\, dy.$$

These are the elements of a symmetric tensor of order two in two dimensions:

$$\int_D \begin{Bmatrix} x \\ y \end{Bmatrix} \{x \quad y\}\, dx\, dy \equiv \int_D \begin{pmatrix} x^2 & xy \\ yx & y^2 \end{pmatrix} dx\, dy.$$

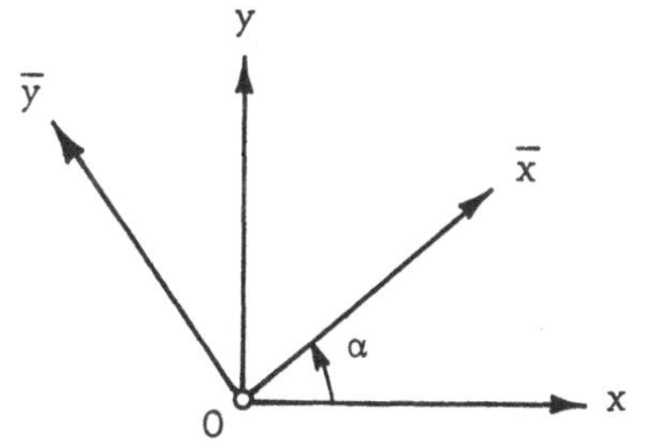


Figure 6.2

The rotation of axes

$$\begin{Bmatrix} \bar{x} \\ \bar{y} \end{Bmatrix} = R \begin{Bmatrix} x \\ y \end{Bmatrix} \quad \text{where} \quad R = \begin{pmatrix} \cos\alpha & \sin\alpha \\ -\sin\alpha & \cos\alpha \end{pmatrix}$$

yields the new matrix of the tensor

$$\int_D \begin{Bmatrix} \bar{x} \\ \bar{y} \end{Bmatrix} \{\bar{x} \quad \bar{y}\} d\bar{x}\, d\bar{y} = R\left[\int_D \begin{Bmatrix} x \\ y \end{Bmatrix} \{x \quad y\} dx\, dy\right] R^T,$$

or, explicitly,

$$I_{\overline{xx}} = I_{xx}\cos^2\alpha + 2I_{xy}\sin\alpha\cos\alpha + I_{yy}\sin^2\alpha,$$

$$I_{\overline{xy}} = (I_{yy} - I_{xx})\sin\alpha\cos\alpha + I_{xy}(\cos^2\alpha - \sin^2\alpha), \tag{6.9}$$

$$I_{\overline{yy}} = I_{xx}\sin^2\alpha - 2I_{xy}\sin\alpha\cos\alpha + I_{yy}\cos^2\alpha.$$

Wherever the origin 0 is chosen, there exist at least two orthogonal principal directions for which the product of inertia $I_{\overline{xy}}$ vanishes. From the second of (6.9), they are seen to correspond to angles $\alpha$ defined by

$$\tan 2\alpha = \frac{2I_{xy}}{I_{xx} - I_{yy}}$$

or

$$\tan \alpha = \frac{I_{yy} - I_{xx} \pm \sqrt{(I_{yy} - I_{xx})^2 + 4I_{xy}^2}}{2I_{xy}} .$$

If $I_{xy} = 0$ then the angles are $\alpha = 0$ and $\alpha = \frac{\pi}{2}$. If also $I_{xx} = I_{yy}$ then all directions are principal and the moments of inertia are the same with respect to every axis passing through 0.

For an arbitrary scalar $\lambda$,

$$\int_D (x + \lambda y)^2 dx\, dy \equiv I_{xx} + 2\lambda\, I_{xy} + \lambda^2\, I_{yy} > 0.$$

Since this quadratic in $\lambda$ can have no real roots, we find that the second invariant of the tensor, which is proportional to the negative of the discriminant, is positive:

$$I_2 = I_{xx}\, I_{yy} - I_{xy}^2 > 0. \tag{6.10}$$

The first invariant, or *polar moment of inertia,* is obviously also positive:

$$I_p = I_{xx} + I_{yy} = \int_D (x^2 + y^2) dx\, dy > 0. \tag{6.11}$$

Now suppose the chosen point 0 is the centroid and calculate the moments of inertia with respect to axes parallel to those of the frame but centered at $x = a$ and $y = b$. The matrix of the tensor will be given by

$$\int_D \begin{Bmatrix} x-a \\ x-y \end{Bmatrix} \{x-a \quad y-b\}\ dx\ dy$$

and, because of the properties (6.7) of the centroid, we find the moments to be

$$I_{xx} + a^2\Omega, \qquad I_{xy} + ab\Omega, \qquad I_{yy} + b^2\Omega.$$

From these it is easy to see that the first and second invariants of the tensor of the moments of inertia are least when the axes issue from the centroid.

## 6.4. The Semi-Inverse Method of Saint-Venant

In accord with the use of the coordinates $(x,y,z)$, we write the stress and strain tensors as follows:

$$\begin{bmatrix} \varepsilon_x & \gamma_{xy}/2 & \gamma_{xz}/2 \\ \gamma_{yx}/2 & \varepsilon_y & \gamma_{yz}/2 \\ \gamma_{zx}/2 & \gamma_{zy}/2 & \varepsilon_z \end{bmatrix} \quad \text{and} \quad \begin{bmatrix} \sigma_x & \tau_{xy} & \tau_{xz} \\ \tau_{yx} & \sigma_y & \tau_{yz} \\ \tau_{zx} & \tau_{zy} & \sigma_z \end{bmatrix}.$$

The *semi-inverse* method of *Saint-Venant* consists in setting

$$\sigma_x \equiv 0, \qquad \sigma_y \equiv 0, \qquad \tau_{xy} \equiv 0, \tag{6.12}$$

and seeking the exact general solution for the equations of elasticity for an isotropic beam under these conditions. In this chapter we develop a solution corresponding to a certain idea of the transmission of stress among fibers parallel to the generators; namely, that they affect each other only tangentially, while they remain oriented parallel to the generators.

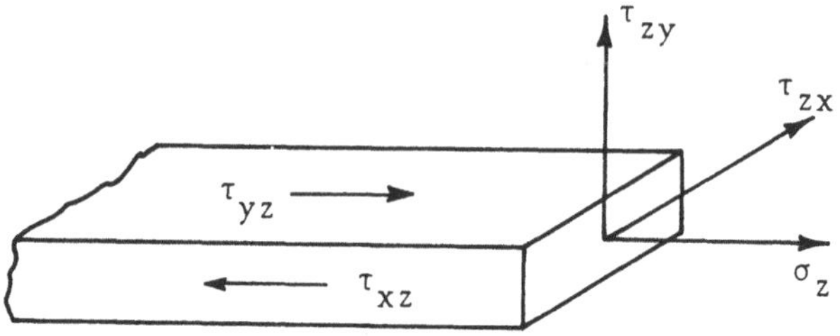


Figure 6.3

In view of the hypotheses (6.12), neglecting the body forces reduces the equilibrium equations to

$$\partial\tau_{zx}/\partial z = 0, \quad \partial\tau_{zy}/\partial z = 0, \tag{6.13}$$

$$\partial\tau_{zx}/\partial x + \partial\tau_{zy}/\partial y + \partial\sigma_z/\partial z = 0. \tag{6.14}$$

Assuming the exterior surface and the cavities free of all traction corresponds to the conditions

$$n_x\sigma_x + n_y\tau_{yx} + n_z\tau_{zx} = 0,$$
$$n_x\tau_{xy} + n_y\sigma_y + n_z\tau_{zy} = 0,$$
$$n_x\tau_{xz} + n_y\tau_{yz} + n_z\sigma_z = 0.$$

Since every normal to the cylindrical surfaces is orthogonal to the axis Oz, $n_z = 0$ and the first two conditions are met trivially by the hypotheses (6.12). The third reduces to

$$n_x\tau_{xz} + n_y\tau_{yz} \equiv \cos\theta\,\tau_{xz} + \sin\theta\,\tau_{yz} = 0 \text{ on } c_i, \tag{6.15}$$

or by using (6.1),

$$\tau_{xz}\,\partial y/\partial s - \tau_{yz}\,\partial x/\partial s = 0. \tag{6.16}$$

This form does not require specification of the direction of passage because it is homogeneous and $\partial y/\partial s$ and $\partial x/\partial s$ change signs simultaneously when the direction is reversed.

The assumption of isotropy leads finally to the constitutive relations, which when simplified by (6.12) become

$$E\varepsilon_x = E\,\partial u/\partial x = -\nu\sigma_z, \quad E\varepsilon_y = E\,\partial v/\partial y = -\nu\sigma_z, \tag{6.17}$$

$$E\varepsilon_z = E\,\partial w/\partial z = \sigma_z, \tag{6.18}$$

$$G\gamma_{xy} = 0 \Rightarrow \partial u/\partial y + \partial v/\partial x = 0, \tag{6.19}$$

$$G\gamma_{xz} = G(\partial u/\partial z + \partial w/\partial x) = \tau_{xz}, \quad G\gamma_{yz} = G(\partial v/\partial z + \partial w/\partial y) = \tau_{yz}. \tag{6.20}$$

Elimination of the displacements from these equations, followed by a simplification from the equilibrium equations, yields the Beltrami-Michell equations

$$\partial^2\sigma_z/\partial x^2 = \partial^2\sigma_z/\partial x\partial y = \partial^2\sigma_z/\partial y^2 = \partial^2\sigma_z/\partial z^2 = 0, \tag{6.21}$$

$$(1 + \nu)\nabla^2\tau_{xz} = -\,\partial^2\sigma_z/\partial x\partial z, \quad (1 + \nu)\nabla^2\tau_{yz} = -\partial^2\sigma_z/\partial y\partial z \tag{6.22}$$

with the notation $\nabla^2 = \partial^2/\partial x^2 + \partial^2/\partial y^2$ for Laplace's operator in two dimensions.

## 6.5. Resultants of Stresses on a Cross Section

Equations (6.21) show that the normal stress $\sigma_z$ in the fibers is distributed according to a linear law of the type

$$\sigma_z = E(\varepsilon + xK_x + yK_y) \tag{6.23}$$

in which the coefficients $\varepsilon$, $K_x$ and $K_y$ are at most linear functions of $z$. These coefficients may be expressed as functions of the resultant *axial force* $T_z$ and the two resultant moments $M_x$, $M_y$ *(bending moments)* acting in a cross section $z$ = constant. We assume that the axis $Oz$ passes through the centroid of each cross section and refer to it as the *centroi-*

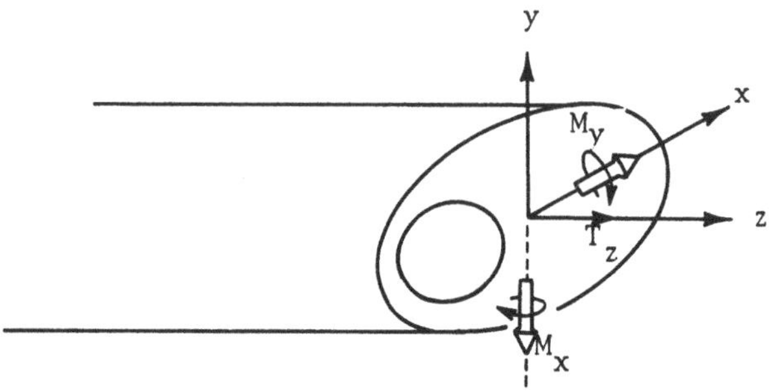


Figure 6.4

*dal fiber.* By using the properties (6.7) we then obtain

$$\begin{aligned} T_z &\equiv \int_D \sigma_z \, dx\, dy = E\Omega\varepsilon, \\ M_x &\equiv \int_D x\sigma_z \, dx\, dy = E(I_{xx}K_x + I_{xy}K_y), \\ M_y &\equiv \int_D y\sigma_z \, dx\, dy = E(I_{xy}K_x + I_{yy}K_y). \end{aligned} \tag{6.24}$$

When the axes are oriented so that the product of inertia vanishes, the equations are uncoupled and their solution is

$$\varepsilon = T_z/E\Omega, \quad K_x = M_x/EI_{xx}, \quad K_y = M_y/EI_{yy}. \tag{6.25}$$

In practice, the principal orientations may be used effectively only if they are known in advance, as for example when the cross section is bilaterally symmetric. If the orientations are not principal, we use the more general formulas

$$K_x = \frac{I_{yy}M_x - I_{xy}M_y}{E(I_{xx}I_{yy} - I_{xy}^2)}, \quad K_y = \frac{-I_{xy}M_x + I_{xx}M_y}{E(I_{xx}I_{yy} - I_{xy}^2)}. \tag{6.26}$$

Substitution of such formulas into (6.23) expresses the normal stress as a function of the axial force and the bending moments.

A formal calculation of the strength of variation of the resultants (6.24) with position along the axis may be started from equation (6.14) for longitudinal equilibrium. For example,

$$\frac{dT_z}{dz} = \int_D \frac{\partial \sigma_z}{\partial z}\, dxdy = -\int_D \left(\frac{\partial \tau_{xz}}{\partial x} + \frac{\partial \tau_{yz}}{\partial y}\right) dxdy$$
$$= \sum_0^n \int_{C_i} (n_x \tau_{xz} + n_y \tau_{yz})\, ds.$$

In the integration we have used the continuity of the shear stresses on the bridges connecting the directrices. In view of the boundary conditions (6.15) this becomes

$$dT_z/dz = 0. \tag{6.27}$$

In the same way one reaches the equations

$$dM_x/dz = T_x, \quad dM_y/dz = T_y, \tag{6.28}$$

where the right hand sides are the *shear forces*

$$T_x = \int_D \tau_{xz}\, dxdy, \quad T_y = \int_D \tau_{yz}\, dxdy. \tag{6.29}$$

For the first of these, for example,

$$\frac{dM_x}{dz} = \int_D x\, \frac{\partial \sigma_z}{\partial z}\, dxdy = -\int_D x\left(\frac{\partial \tau_{xz}}{\partial x} + \frac{\partial \tau_{yz}}{\partial y}\right) dx\, dy$$
$$= -\sum_0^n \int_{C_i} x(n_x \tau_{xz} + n_y \tau_{yz})\, ds + \int_D \tau_{xz}\, dxdy$$

and the result follows by use of the boundary conditions.

It is obvious from (6.13) that the shear forces, like

the shear stresses themselves, are independent of the coordinate $z$:

$$dT_x/dz = 0, \qquad dT_y/dz = 0. \tag{6.30}$$

Like $K_x$ and $K_y$, then, the bending moments are linear functions of $z$.

The derivatives of $K_x$ and $K_y$ are two essential parameters in the development of the theory:

$$a \equiv \frac{dK_x}{dz} = \frac{I_{yy}T_x - I_{xy}T_y}{E(I_{xx}I_{yy} - I_{xy}^2)}, \quad b \equiv \frac{dK_y}{dz} = \frac{-I_{xy}T_x + I_{yy}T_y}{E(I_{xx}I_{yy} - I_{xy}^2)}. \tag{6.31}$$

Evidently, the derivatives $a$ and $b$ are linear functions of the shear forces.

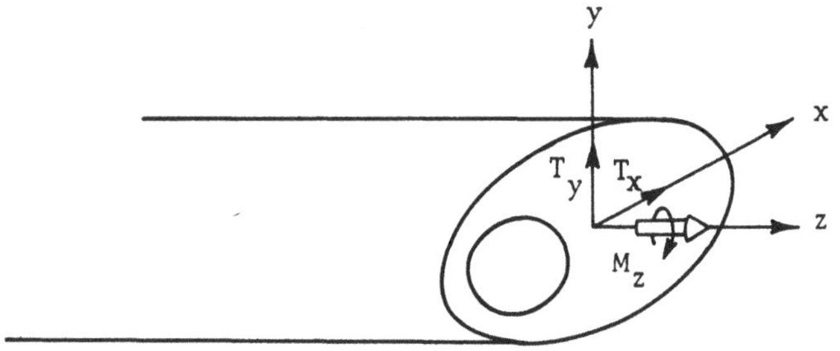


Figure 6.5

The last resultant in the cross section is the *twisting moment*

$$M_z = \int_D (x\tau_{yz} - y\tau_{xz})dx\,dy. \tag{6.32}$$

It is also independent of $z$, i.e.,

$$dM_z/dz = 0. \tag{6.33}$$

Equations (6.28) and the constancy of the shear forces,

the axial force, and the twisting moment are also the expression of the global equilibrium equations for a segment of the beam of length $dz$.

## 6.6. Calculation of the Transverse Displacements

Equations (6.17) and (6.19) show that the transverse displacements obey the Cauchy-Riemann equations

$$\partial u/\partial x = \partial v/\partial y, \qquad \partial u/\partial y = -\partial v/\partial x, \tag{6.34}$$

which imply that the complex form $u + iv$ is an analytic function of the complex variable $x + iy$. By setting

$$u + iv = u_0 + iv_0 + (\rho+i\omega_0)(x+iy) + (u_2+iv_2)(x+iy)^2,$$

where the coefficients are at most functions of $z$, we obtain

$$\partial u/\partial x = \partial v/\partial y = \rho + 2u_2x - 2v_2y.$$

When this result is substituted into (6.17) and $\sigma_z$ is replaced by its expression (6.23), comparison yields

$$\rho = -\nu\varepsilon, \qquad u_2 = -\frac{\nu}{2}K_x, \qquad v_2 = \frac{\nu}{2}K_y.$$

Thus we find

$$\begin{aligned} u &= u_0(z)-y\omega_0(z) - \nu\{\varepsilon x + \tfrac{1}{2}(x^2-y^2)K_x(z) + xyK_y(z)\}, \\ v &= v_0(z)+x\omega_0(z) - \nu\{\varepsilon y + xy\,K_x(z) - \tfrac{1}{2}(x^2-y^2)K_y(z)\}. \end{aligned} \tag{6.35}$$

By setting $x = 0$ and $y = 0$, we see that $u_0(z)$ and $v_0(z)$ represent the transverse displacements of the centroidal fiber. The material rotation of a fiber about its axis is

$$\omega_z = \frac{1}{2}\left(\frac{\partial v}{\partial x} - \frac{\partial u}{\partial y}\right) = \omega_0(z) - \nu\{y\,K_x(z) - x\,K_y(z)\}. \tag{6.36}$$

From this formula it follows that $\omega_o(z)$ is the rotation of the centroidal fiber. The other terms in (6.35) are Poisson effects resulting from transverse contractions or expansions of the fibers caused by their state of longitudinal tension or compression.

Differentiating equations (6.20) with respect to $z$ and using (6.13) and (6.18) yields

$$\begin{aligned} \frac{\partial^2 u}{\partial z^2} &= -\frac{\partial^2 w}{\partial x \partial z} = -\frac{1}{E}\frac{\partial \sigma_z}{\partial x} = -K_x(z), \\ \frac{\partial^2 v}{\partial z^2} &= -\frac{\partial^2 w}{\partial y \partial z} = -\frac{1}{E}\frac{\partial \sigma_z}{\partial y} = -K_y(z), \end{aligned} \tag{6.37}$$

which allow $K_x$ and $K_y$ to be interpreted as the linearized curvatures of the fibers, and in particular those of the centroidal fiber:

$$d^2u_o/dz^2 = -K_x, \qquad d^2v_o/dz^2 = -K_y. \qquad (6.37)'$$

By differentiating these formulas again we find a kinematic interpretation of the parameters $a$ and $b$ as derivatives of curvatures:

$$d^3u_o/dz^3 = -a, \qquad d^3v_o/dz^3 = -b. \tag{6.38}$$

Now differentiate formulas (6.35) twice with respect to $z$. Poisson's effects disappear, being only linear in $z$, and comparison of the result with (6.37) yields

$$d^2\omega_o/dz^2 = 0.$$

The constant parameter

$$\theta = d\omega_o/dz, \tag{6.39}$$

or *twist* of the centroidal fiber, plays, along with $a$ and $b$, an essential part in the calculation of the shear stresses. The kinematic variables $a$ and $b$ are related in a simple way to the shear forces by the stiffness relations

$$T_x = EI_{xx}a + EI_{xy}b, \qquad T_y = EI_{xy}a + EI_{yy}b, \tag{6.40}$$

obtained by inverting (6.31). On the other hand, the relation between $\theta$ and the resultants of the shear stresses is more involved and its development is postponed until Section 6.13.

We observe finally that the twist of a general fiber

$$\partial\omega_z/\partial z = \theta + \nu(bx - ay), \tag{6.41}$$

unlike that of the centroidal fiber, exhibits a Poisson effect caused by the linearly varying shear forces.

In order to facilitate the evaluation of the integrals related to the transverse displacements, it will help to put the latter into the form

$$u = u_o(z) - \partial\phi/\partial x - \partial\psi/\partial y, \quad v = v_o(z) - \partial\phi/\partial y + \partial\psi/\partial x, \tag{6.42}$$

with

$$\begin{aligned} \phi &= \nu\left(\varepsilon\,\frac{x^2+y^2}{2} + \frac{x^3}{3}K_x + \frac{y^3}{3}K_y\right), \\ \psi &= \frac{x^2+y^2}{2}\,\omega_o(z) - \frac{\nu}{6}(y^3+3yx^2)K_x + \frac{\nu}{6}(x^3+3xy^2)K_y. \end{aligned} \tag{6.43}$$

The derivatives of these relations with respect to $z$ may similarly be written as

$$\frac{\partial u}{\partial z} = \frac{du_o}{dz} - \frac{\partial\lambda}{\partial x} - \frac{\partial\mu}{\partial y}, \quad \frac{\partial v}{\partial z} = \frac{dv_o}{dz} - \frac{\partial\lambda}{\partial y} + \frac{\partial\mu}{\partial x}, \tag{6.42$'$}$$

with

$$\begin{aligned} \lambda &\equiv \frac{\partial\phi}{\partial z} = \frac{\nu}{3}(ax^3 + by^3), \\ \mu &\equiv \frac{\partial\psi}{\partial z} = \frac{x^2+y^2}{2}\,\theta - \frac{\nu}{6}(y^3+3yx^2)a + \frac{\nu}{6}(x^3+3xy^2)b. \end{aligned} \tag{6.43$'$}$$

## 6.7. Equations Governing the Shear Stresses

Saint-Venant's assumptions lead to very simple results for the longitudinal stress in the fibers and their transverse displacement. On the other hand, determination of shear stresses and the longitudinal displacement requires the solution of problems involving Laplace's operator in two dimensions.

It is advantageous to regard the pair $(\tau_{xz},\tau_{yz})$ as the components of a vector $\tau$ to be defined at each point of the region D of a cross section. From (6.13) we know already that this vector is independent of the choice of cross section. We may express the divergence and the z-component of the curl of this vector field in terms of the parameters $(a,b,\theta)$:

$$\begin{aligned} \operatorname{div}\tau &= \partial\tau_{xz}/\partial x + \partial\tau_{yz}/\partial y = -\partial\sigma_z/\partial z = -E(ax+by), \\ \operatorname{rot}_z\tau &= \partial\tau_{yz}/\partial x - \partial\tau_{xz}/\partial y, \end{aligned} \tag{6.44}$$

and by (6.20) and (6.41)

$$\operatorname{rot}_z\tau = G\,\frac{\partial}{\partial z}\Big(\frac{\partial v}{\partial x} - \frac{\partial u}{\partial y}\Big) = 2G\,\frac{\partial\omega_z}{\partial z} = 2G\{\theta+\nu(bx-ay)\}. \tag{6.45}$$

The field is uniquely determined by these equations together with the boundary condition (6.16)

$$\tau_{xz}\,\frac{\partial y}{\partial s} - \tau_{yz}\,\frac{\partial x}{\partial s} = 0 \quad \text{on} \quad c_i, \qquad i = 0,1,\ldots n$$

and the circuital conditions

$$\int_{c_i} \tau\cdot ds = 2G\Omega_i\{\theta+\nu(b\hat{x}_i - a\hat{y}_i)\}, \qquad i = 1,2\ldots n, \tag{6.46}$$

where each directrix $c_i$ is traversed in the mathematically positive sense. Conditions (6.46) arise from an interpretation of the global compatibility conditions necessary for the univalence of the longitudinal displacement, and will be derived in the following section. The corresponding local compatibility conditions are guaranteed by satisfying the equilibrium equations and the Beltrami-Michell equations. Now, equations (6.21) have been used for the construction of the field $\sigma_z$ and, as one verifies easily, equations (6.22) are satisfied by (6.44) and (6.45). One may thus consider that all the integrability conditions for the remaining displacement $w$ are satisfied.

We finally reduce the determination of the shear stresses to Poisson problems, with boundary conditions of Dirichlet or Neumann type, by writing

$$\tau_{xz} = E\frac{\partial\Phi}{\partial x} + G\frac{\partial\Psi}{\partial y}, \quad \tau_{yz} = E\frac{\partial\Phi}{\partial y} - G\frac{\partial\Psi}{\partial x}. \tag{6.47}$$

Equations (6.44), (6.45), (6.16), and (6.46) thus become

$$\nabla^2\Phi = -(ax + by) \quad \text{with} \quad \frac{\partial\Phi}{\partial n} = 0 \quad \text{on} \quad c_i, \; i = 0,1,\dots,n;$$
$$\nabla^2\Psi = -2\theta - 2\nu(bx - ay) \quad \text{with } d\Psi = 0 \text{ on } c_i, \tag{6.48}$$

$$\int_{c_i} \frac{\partial\Psi}{\partial n}\, ds = 2\Omega_i\theta + 2\nu\Omega_i(b\hat{x}_i - a\hat{y}_i), \; i = 0,1,\dots,n. \tag{6.49}$$

Since the addition of a constant to $\Psi$ alters neither the stress field nor the equations to be satisfied it is permissible to put

$$\Psi = 0 \quad \text{on} \quad c_0, \; \Psi = \Psi_i \quad \text{on} \quad c_i, \quad i = 1,2,\dots,n, \qquad (6.49)'$$

and the $n$ constant values taken by the function on the directrices of the cavity are to be determined precisely by the $n$ circuital conditions.

## 6.8. Calculation of the Longitudinal Displacement

We may integrate (6.37)' in the form

$$\frac{du_0}{dz} = q - \int_0^z K_x(\zeta)d\zeta, \qquad \frac{dv_0}{dz} = -p - \int_0^z K_y(\zeta)d\zeta, \tag{6.50}$$

where $p$ and $q$ are the slopes of the centroidal fiber at the reference section $z = 0$.

Equations (6.18) and (6.23) yield

$$\partial w/\partial z = \sigma_z/E = \varepsilon + x\, K_x(z) + y\, K_y(z). \tag{6.51}$$

The integral of (6.51) may be written

$$w = \varepsilon z + x\int_0^z K_x(\zeta)d\zeta + y\int_0^z K_y(\zeta)d\zeta + w_0 + py - qx + g(x,y). \tag{6.52}$$

The first three terms form a particular integral, vanishing at the reference section $z = 0$. From the general integral of the equation with vanishing right hand side, which is an arbitrary function of $(x,y)$, we have explicitly extracted a linear form in such a way that the parameters $(p,q)$ now appear as small rotations of the beam about the axes $Ox$ and $Oy$. With these and the parameter $w_0$, which is a longitudinal translation, we have exhibited three of the six degrees of freedom as a rigid body which always arise when strains are integrated to give displacements. The three others come from the constants of integration of equations (6.39) (rotation about $Oz$) and (6.50) (lateral translations).

It remains to determine the function $g(x,y)$, called the *warping* of the cross section because, apart from certain special cases, it is not linear in the variables $x$ and $y$ as are the other terms in $w$, and hence does not preserve the flatness of the section. We shall see that $g(x,y)$ is defined only up to an additive constant, which will be $w_0$ if one makes $g(x,y)$ unique by the supplementary condition

$$\int_D g \, dx \, dy = 0. \tag{6.53}$$

The longitudinal displacement in the reference section is

$$w(x,y,0) = py - qx + w_0 + g(x,y), \tag{6.54}$$

and if desired one could adjust the parameters of the linear form in such a way that $w(x,y,0)$ would become a *true warping*, enjoying the properties

$$\int_D w(x,y,0)dxdy = \int_D w(x,y,0)xdxdy = \int_D w(x,y,0)ydxdy = 0. \tag{6.55}$$

In fact, by virtue of (6.7) and (6.8), (6.53)-(6.55) imply

$$w_0 = 0, \quad I_{xx}q - I_{xy}p = \int_D gx \, dxdy,$$
$$I_{xy}q - I_{yy}p = \int_D gy \, dxdy. \tag{6.56}$$

We notice that the latter two of (6.56) do not require that $g$ necessarily satisfy (6.53).

Equation (6.54) shows that if Saint-Venant's hypotheses are not to be violated, the longitudinal displacement $w(x,y,0)$ at the reference section may not be specified arbitrarily. For example, if this section were a perfectly clamped end, one would wish to have $w(x,y,0) \equiv 0$, but in general this would be impossible in the context of Saint-Venant's theory. However, as noted by *Weinstein*, the clamping can be approximated by requiring $w(x,y,0)$ to be a true warping. The approximation is in the sense of least squares, i.e., the integral $\int_D w^2(x,y,0)dxdy$ is minimized if equations (6.56) are satisfied.

Another way of deriving (6.56) as approximate conditions for clamping is to suppose that a corrective solution, with stresses

$$\hat{\sigma}_z = f_0(py - qx + w_0 + g)$$

proportional to (6.54), could negate the displacement of (6.54). Since these additional stresses should be statically equivalent to zero, they satisfy the conditions

$$\int_D \hat{\sigma}_z dxdy = \int_D \hat{\sigma}_z \, x \, dxdy = \int_D \hat{\sigma}_z \, y \, dxdy = 0$$

which are again equivalent to the equations (6.56).

We now calculate the shear strains from equations (6.35) and (6.42)' with (6.50) and (6.52), getting

$$\gamma_{xz} = -\frac{\partial\lambda}{\partial x} - \frac{\partial\mu}{\partial y} + \frac{\partial g}{\partial x}, \quad \gamma_{yz} = -\frac{\partial\lambda}{\partial y} + \frac{\partial\mu}{\partial x} + \frac{\partial g}{\partial y}\,. \tag{6.57}$$

Because the derivatives of the functions $\lambda$ and $\mu$ vanish for $x = 0$ and $y = 0$, one sees that in the absence of warping the shear strains vanish at the level of the centroidal fiber, the sections thus remaining locally plane and orthogonal to this fiber.

We have seen that all the *local* conditions for compatibility were satisfied for the integration of $w$ and then of $g$. In fact, elimination of $g$ among equations (6.57) recovers (6.45). According to the general theory, the *global* compatibility conditions

$$\oint dg = \oint \frac{\partial g}{\partial x} dx + \frac{\partial g}{\partial y} dy = 0,$$

or conditions for the univalence of $g$, only need to be satisfied for one member of each of the equivalence classes of irreducible circuits. For these we choose the directrices of the cavities, so we have

$$\int_{c_i} \gamma_{xz}\, dx + \gamma_{yz}\, dy + \int_{c_i} \frac{\partial\mu}{\partial y}\, dx - \frac{\partial\mu}{\partial x}\, dy = 0$$

because

$$\int_{c_i} \frac{\partial\lambda}{\partial x}\, dx + \frac{\partial\lambda}{\partial y}\, dy = \int_{c_i} d\lambda = 0.$$

With the constitutive equations (6.20), this becomes

$$\frac{1}{G}\int_{c_i} \tau\cdot ds = \theta \int_{c_i} xdy - ydx - \frac{\nu a}{2}\int_{c_i} (x^2-y^2)dx + 2xy\, dy$$
$$- \frac{\nu b}{2}\int_{c_i} 2xydx - (x^2-y^2)dy.$$

One may therefore use in the right-hand side the formulas (6.5)' and (6.6)' for each circuit being traversed in the direct sense. Thus one reaches the justification for the proposed circuital conditions (6.46).

Another expression for the shear strains results from formulas (6.47):

$$\gamma_{xz} = 2(1+\nu)\frac{\partial\Phi}{\partial x} + \frac{\partial\psi}{\partial y}, \quad \gamma_{yz} = 2(1+\nu)\frac{\partial\Phi}{\partial y} - \frac{\partial\psi}{\partial x}. \tag{6.58}$$

On comparing this with (6.57) one verifies that the Cauchy-Riemann relations

$$\partial H/\partial x = \partial Z/\partial y, \quad \partial H/\partial y = -\partial Z/\partial x \tag{6.59}$$

are satisfied by the functions

$$H = g - 2(1+\nu)\Phi - \lambda \tag{6.60}$$

and

$$Z = \Psi + \mu. \tag{6.61}$$

The harmonic function $Z$ is subject to the boundary conditions of Dirichlet type $dZ = d\mu$ on $c_i$, $i = 0,1,\dots,n$, or

$$\begin{aligned} Z &= \mu \quad \text{on} \quad c_o, \\ Z &= \mu + \psi_i \quad \text{on} \quad c_i, \qquad i = 1,2,\ldots,n. \end{aligned} \tag{6.62}$$

In addition, since the Cauchy-Riemann structure gives on the boundaries the relations $\partial H/\partial n = \partial Z/\partial s$, $\partial H/\partial s = -\partial Z/\partial n$, the circuital conditions are homogeneous for $Z$:

$$\int_{c_i} \frac{\partial Z}{\partial n}\, ds = -\int_{c_i} \frac{\partial H}{\partial s}\, ds = 0, \qquad i = 1,2,\ldots,n. \tag{6.62)'$$

In the same way, Neumann's data for the conjugate harmonic function $H$ are supplied directly by the derivatives of Dirichlet's data for $Z$:

$$\partial H/\partial n = \partial\mu/\partial s \quad \text{on} \quad c_i, \qquad i = 0,1,\ldots,n. \tag{6.63}$$

One sees in particular from (6.60) that the warping solves the Poisson problem with Neumann's data

$$\nabla^2 g = -2(ax+by), \quad \partial g/\partial n = \partial\lambda/\partial n + \partial\mu/\partial s \quad \text{on } c_i, \quad i = 0,1,\ldots,n, \tag{6.64}$$

and is determined only up to an additive constant. If this problem is solved the shear stresses may be derived from it by multiplying equations (6.57) by $G$.

In an equivalent way, one may set

$$g = k + \lambda$$

and solve the problem

$$\nabla^2 k = -2(1+\nu)(ax+by), \quad \partial k/\partial n = \partial\mu/\partial s \text{ on } c_i,\ i = 0,1,\ldots,n,$$

the stresses being then given by

$$\tau_{xz} = G(\partial k/\partial x - \partial\mu/\partial y), \quad \tau_{yz} = G(\partial k/\partial y + \partial\mu/\partial x).$$

Consider now the conditions under which the cross section can remain plane. In view of (6.52) this would require $g(x,y) = \alpha x + \beta y$. The Laplacian of this function being zero, one concludes immediately that $a = 0$ and $b = 0$; i.e., the presence of a shear force necessarily destroys the planeness of a cross section. The possibility $\theta \neq 0$ remains, but its viability rests on verification of the boundary conditions

$$\alpha \frac{\partial x}{\partial n} + \beta \frac{\partial y}{\partial n} = \alpha \frac{\partial y}{\partial s} - \beta \frac{\partial x}{\partial s} = \frac{\partial}{\partial s}(\alpha y - \beta x) = \frac{\partial}{\partial s}\{\frac{\theta}{2}(x^2+y^2)\}.$$

To comply, a directrix must have an equation of the form

$$\frac{1}{2}\theta (x^2 + y^2) = \alpha y - \beta x + \gamma_i, \quad i = 0,1,\ldots,n.$$

The directrices are thus concentric circles, implying that there can be no more than one cavity. In order that the axes have their origin at the centroid, moreover, it is necessary to take $\alpha = 0$ and $\beta = 0$, and then $g \equiv 0$. For a single circle we have the case of the circular bar in torsion; for two concentric circles, the case of the circular tube.

## 6.9. Separation of Solutions

We have formulated the general solution of the problem of prismatic beams in the context of Saint-Venant's hypotheses. Because of linearity, the solution may be regarded as a superposition of several partial solutions. The simplest ones are those for which there are no shear stresses, i.e., those corresponding to vanishing parameters $\theta$, a, and b.

(i) Extension. All the fibers are subjected to the same normal stress $\sigma_z = T_z/\Omega$ and suffer the same specific elongation $\varepsilon$. The beam is a state of uniaxial stress and the global stiffness relation is $T_z = E\Omega\varepsilon$. Due to Poisson's effect there is a transverse contraction given by $u = -\nu\varepsilon x$, $v = -\nu\varepsilon y$.

(ii) Pure bending. Since the shear forces vanish, the beam is subjected to constant bending moments and the normal stress in the fibers is distributed linearly according to the relation $\sigma_z = xK_x + yK_y$, where the constant curvatures are connected with the bending moments by the stiffness equations

$$M_x = EI_{xx}K_x + EI_{xy}K_y, \qquad M_y = EI_{xy}K_x + EI_{yy}K_y.$$

The absence of shear stresses implies the relations

$$\partial u/\partial z = -\partial w/\partial x, \qquad \partial v/\partial z = -\partial w/\partial y,$$

showing that all fibers remain orthogonal to the cross sections. As in the case of extension, however, they suffer a distortion in their plane by Poisson's effect:

$$u = -\nu\,\tfrac{1}{2}(x^2 - y^2)K_x - \nu\,xyK_y,$$
$$v = -\nu\,xyK_x + \nu\,\tfrac{1}{2}(x^2 - y^2)K_y.$$

Although the sections remain plane, this distortion does not permit consideration of a perfectly clamped end without introducing a correction to Saint-Venant's solution.

The partial solutions for extension and pure bending are characterized by the independence of the longitudinal fibers, in the sense that they exert no force on each other. On the other hand, tangential interaction among fibers plays a fundamental role in torsion and in bending accompanied by shear forces. We will eventually try to separate the contribution of the pure torsion from that of the bending.

## 6.10. Pure Torsion

Pure torsion may be defined unambiguously as the partial solution for which $\sigma_z \equiv 0$, which implies the absence of bending moments and shear forces.

Therefore, from the kinematic point of view,

$$K_x = K_y = a = b = \varepsilon = 0.$$

Here we replace $\theta$ by $\hat{\theta}$ since another contribution to $\theta$ will arise, as we will see, from bending without torsion.

The shear stresses here satisfy the simplified equations

$$\operatorname{div} \tau = 0 \tag{6.44'}$$

$$\operatorname{rot}_z \tau = 2G\hat{\theta}. \tag{6.45'}$$

The resultant of $\tau$ on a cross section will be the twisting moment C. On the directrices the field obeys the homogeneous conditions (6.16) and the simplified circuital conditions

$$\int_{c_i} \tau \cdot ds = 2G\hat{\theta}\Omega_i, \quad i = 1,2,\ldots,n. \tag{6.46'}$$

Condition (6.44)', analogous to the incompressibility condition of two dimensional hydrodynamics, is satisfied by introducing *Prandtl's stress function* $\Theta(x,y)$, analogous to a stream function, such that

$$\tau_{xz} = G\hat{\theta}\,\frac{\partial\Theta}{\partial y}, \quad \tau_{yz} = -G\hat{\theta}\,\frac{\partial\Theta}{\partial x}. \tag{6.47'}$$

This amounts to setting $\Phi \equiv 0$ and $\Psi = \hat{\theta}\Theta(x,y)$ in the general theory. It follows that the stress function is a solution of the problem

$$\begin{aligned} \nabla^2\Theta &= -2 \\ d\Theta &= 0 \quad \text{on } c_i, \quad i = 0,1,\ldots,n. \end{aligned} \tag{6.49'}$$

Because the addition of a constant to $\Theta$ does not alter the stresses, we may set $\Theta = 0$ on $c_0$ and determine the value $\alpha_i$ of $\Theta$ on each interior directrix by the circuital conditions

$$\int_{c_i} \frac{\partial\Theta}{\partial n}\, ds = 2\Omega_i, \qquad i = 1,2,\ldots,n. \tag{6.49)''$$

In order to calculate the twisting moment $C$, one may use formula (6.32) to find the moment with respect to the axis $Oz$ (the resultant force will vanish), obtaining the stiffness formula

$$C = GJ\hat{\theta} \tag{6.65}$$

where $J$ is the *torsional stiffness constant*

$$J = -\int_D \left(x\,\frac{\partial\Theta}{\partial x} + y\,\frac{\partial\Theta}{\partial y}\right) dx\, dy, \tag{6.66}$$

With the directrices of the cavities traversed in the retrograde sense, Green's transformation with $\phi = \Theta$ and $\psi = \frac{1}{2}(x^2 + y^2)$ yields

$$\int_D \left(2\Theta + x\,\frac{\partial\Theta}{\partial x} + y\,\frac{\partial\Theta}{\partial y}\right) dxdy = \sum_0^n \int_{c_i} \Theta\,\left(x\,\frac{\partial x}{\partial n} + y\,\frac{\partial y}{\partial n}\right) ds$$

$$= \sum_1^n \alpha_i \int_{c_i} x\, dy - y\, dx = -2\sum_1^n \alpha_i\Omega_i,$$

Equation (6.66) then becomes

$$J = 2\int_D \Theta\, dxdy + 2\sum_1^n \alpha_i\Omega_i. \tag{6.66)'$$

This formula suggests the definition of a stress function $\hat{\Theta}$ with values extended to the interiors of the cavities by the convention

$$\hat{\Theta} = \Theta \ \text{ in } D, \qquad \hat{\Theta} = \alpha_i \ \text{ in } \Omega_i, \quad i = 1,2,\ldots,n.$$

Using this convention, we may rewrite (6.66)' as

$$J = 2 \int_{\Omega_0} \hat{\theta}\ dx\ dy. \tag{6.66}''$$

The displacements of pure torsion reduce to

$$u = u_0(z) - \omega_0(z)y, \quad v = v_0(z) + \omega_0(z)x, \quad w = p_\theta y - q_\theta x + g_\theta(x,y)$$

where

$$u_0(z) = q_\theta z + u_0(0), \quad v_0(z) = -p_\theta z + v_o(0), \quad \omega_o(z) = \hat{\theta} z + r_\theta.$$

We define a point $(x_F, y_F)$ of the cross section by the relations

$$q_\theta = \hat{\theta}\ y_F, \quad p_\theta = \hat{\theta}\ x_F.$$

The transverse displacements then take the forms

$$u = u_o(0) - yr_\theta - \hat{\theta} z(y - y_F), \qquad v = v_o(0) + xr_\theta + \hat{\theta} z(x - x_F).$$

The first two terms on the right hand sides are small rigid body displacements; the third ones are torsional displacements by which each section rotates about the point $(x_F, y_F)$ without deformation in its plane. This point thus becomes a *center of torsion.* The position of the center of torsion therefore depends on the values of the rotation parameters $(p_\theta, q_\theta)$ defined by the kinematic support of the beam.

The warping displacement being clearly proportional to the twist, we introduce the normalized warping function $H_\theta(x,y)$ by the relation

$$g_\theta(x,y) = \hat{\theta} H_\theta(x,y). \tag{6.67}$$

The notation is consistent, because in equation (6.60) the functions $\Phi$ and $\lambda$ vanish with $a$ and $b$. The normalized warping function thus satisfies the equations

$$\begin{aligned} &\nabla^2 H_\theta = 0, \\ &\frac{\partial H_\theta}{\partial n} = \frac{1}{2}\frac{\partial}{\partial s}(x^2 + y^2) \quad \text{on} \quad c_i, \quad i = 0,1,\dots,n. \end{aligned} \tag{6.68}$$

If this Neumann's problem is solved one can also calculate the stresses from the equations

$$\tau_{xz} = G\hat{\theta}(\partial H_\theta/\partial x - y), \quad \tau_{yz} = G\hat{\theta}(\partial H_\theta/\partial y + x). \tag{6.69}$$

Finally, consider the harmonic conjugate of $H_\theta$

$$Z_\theta = \Theta + \frac{1}{2}(x^2 + y^2), \tag{6.70}$$

which satisfies boundary conditions of Dirichlet type

$$\begin{aligned} &Z_\theta = \frac{1}{2}(x^2 + y^2) \quad \text{on} \quad c_o \\ &Z_\theta = \frac{1}{2}(x^2 + y^2) + \alpha_i \quad \text{on} \quad c_i, \quad i = 1,2,\dots,n, \end{aligned} \tag{6.70'}$$

and the auxiliary relations

$$\int_{c_i} \frac{\partial Z_\theta}{\partial n}\, ds = 0, \quad i = 1,2,\dots,n.$$

The stresses follow from formulas (6.69) with the help of the Cauchy-Riemann equations

$$\partial H_\theta/\partial x = \partial Z_\theta/\partial y, \quad \partial H_\theta/\partial y = -\partial Z_\theta/\partial x.$$

By comparing (6.47)' with (6.69) one concludes that

$$x = -\partial H_\theta/\partial y - \partial\Theta/\partial x, \quad y = \partial H_\theta/\partial x - \partial\Theta/\partial y.$$

We now calculate the polar moment inertia of the section, getting

$$\begin{aligned} I_p = \int_D (x^2+y^2)\,dxdy = &\int_D \left\{\left(\frac{\partial H_\theta}{\partial x}\right)^2 + \left(\frac{\partial H_\theta}{\partial y}\right)^2\right\} dxdy \\ &+ \int_D \left\{\left(\frac{\partial\Theta}{\partial x}\right)^2 + \left(\frac{\partial\Theta}{\partial y}\right)^2\right\} dxdy \end{aligned} \tag{6.71}$$

because, by Stokes' transformation,

$$\int_D \left(\frac{\partial H_\theta}{\partial y}\frac{\partial \Theta}{\partial x} - \frac{\partial H_\theta}{\partial x}\frac{\partial \Theta}{\partial y}\right) dxdy = -\sum_0^n \int_{C_i} H_\theta \, d\Theta = 0.$$

On the other hand, by Green's first formula applied to $\Theta$ grad $\Theta$,

$$\int_D \text{grad}\,\Theta \cdot \text{grad}\,\Theta \, dxdy = -\int_D \Theta\nabla^2\Theta \, dxdy + \sum_0^n \int_{C_i} \Theta \frac{\partial \Theta}{\partial n} ds$$

$$= 2\int_D \Theta \, dxdy + 2\sum_1^n \alpha_i\Omega_i = J. \qquad (6.72)$$

Substitution of this result in (6.71) yields

$$J = I_p - \int_D \left\{\left(\frac{\partial H_\theta}{\partial x}\right)^2 + \left(\frac{\partial H_\theta}{\partial y}\right)^2\right\} dxdy, \qquad (6.73)$$

displaying the contribution of warping to the torsional rigidity. It is always negative and therefore the torsion constant $J$ in Saint-Venant's theory is always less than the polar moment of inertia. It is equal to it for a circular bar or tube, wherein the warping vanishes.

## 6.11. The Center of Torsion for a Fully Constrained Section

By using Weinstein's approximate condition for full constraint against warping

$$\int_D w^2(x,y,0)\,dxdy \quad \text{minimum}$$

we find here, as a particular case of equations (6.56) with the interpretation (6.67),

$$\begin{aligned} I_{xx}y_F - I_{xy}x_F &= \int_D H_\theta x \, dxdy \\ I_{xy}y_F - I_{yy}x_F &= \int_D H_\theta y \, dxdy. \end{aligned} \qquad (6.74)$$

This definition of the center of torsion had already been proposed by *Kappus* prior to Weinstein.

## 6.12. Bending without Torsion

We are concerned here with bending in the presence of shear forces. At the outset the definition of bending without torsion seems inevitably bound to involve a convention. Indeed, from a kinematic viewpoint equation (6.41) shows that as a consequence of Poisson's effect it would be impossible to annul the twist in all the fibers. The convention $\theta = 0$ annuls the twist of the centroidal fiber or, as is more significant, the average over the section of the twist of all the fibers:

$$\int_D \frac{\partial \omega_z}{\partial z}\, dx\, dy = 0.$$

Such a definition of bending without torsion was proposed by *Timoshenko* and followed by several other authors including *Sokolnikoff*, *Mindlin*, and *Salvadori*. Its major defect is the failure to separate the strain energy into parts associated with bending and pure torsion.

Another way to adopt a convention is to choose as the center of bending the point in a cross section through which pass the lines of action of the resultant shear forces. Saint-Venant himself had chosen to study the case in which the resultants pass through the centroid (here the origin of the axes). More logically, however, one could choose the case in which the center coincides with the center of torsion in the sense of Kappus and Weinstein. It happens that the last definition of bending without torsion coincides with the definition proposed by Trefftz, based on the condition that the strain energy in pure torsion and in bending without torsion should be uncoupled. There follow certain simplifications with respect to Timoshenko's definition; the now common center of bending and torsion has coordinates independent of Poisson's ratio and the shear stresses induced by Poisson's effect becomes statically equivalent to zero. Finally, this definition can be reconciled with the kinematic viewpoint; it annuls a weighted mean of the twist of the fibers, the weighting function being none other than Prandtl's stress function.

Since the twist of the fibers here is due only to a Poisson's effect, we set $\theta = \nu c$ in (6.41), obtaining

$$\partial\omega_z/\partial z = \nu(bx - ay + c) \tag{6.75}$$

and try later to determine the parameter $c$ so as to uncouple the strain energies as just discussed.

The equations governing the shear stresses here are

$$\operatorname{div} \tau = -E(ax + by), \quad \operatorname{rot}_z \tau = 2\nu G(bx - ay + c).$$

We now introduce a solution of the type (6.47) as proposed by *Weber*, but to distinguish from the case of pure torsion,we replace $\Psi$ by $K$. We thus arrive at problem (6.48), i.e.,

$$\nabla^2\Phi = -(ax + by) \quad \text{with} \quad \partial\Phi/\partial n = 0 \quad \text{on} \quad c_i, \quad i = 0,1,\dots,n,$$

and the special case of (6.49)

$$\nabla^2 K = -2\nu(bx-ay+c) \quad \text{with} \quad dK = 0 \quad \text{on} \quad c_i, \; i = 0,1,\dots,n.$$

By reasoning as before, the boundary conditions become

$$K = 0 \text{ on } c_o; \qquad K = \beta_i \text{ on } c_i, \qquad i = 1,2,\dots,n. \tag{6.76}$$

The constants $\beta_i$ are to be determined by the circuital conditions

$$\int_{c_i} \frac{\partial K}{\partial n}\, ds = 2\nu\Omega_i(b\hat{x}_i - a\hat{y}_i + c). \tag{6.76$'$}$$

The field generated by the potential $\Phi$ will be called the *principal field*; that generated by the function $K$, which takes account of Poisson's effects, will be the *secondary field*. In order to define the secondary field it is necessary to determine the parameter $c$. To this end we calculate the strain energy density, which according to Clapeyron's interior theorem is given by

$$W = \frac{1}{2}(\varepsilon_x\sigma_x + \varepsilon_y\sigma_y + \varepsilon_z\sigma_z + \tau_{yz}\gamma_{yz} + \tau_{zx}\gamma_{zx} + \tau_{xy}\gamma_{xy}),$$

and which is already reduced by the hypotheses of the semi-inverse method to

$$W = \frac{1}{2}(\varepsilon_z\sigma_z + \tau_{yz}\gamma_{yz} + \tau_{zx}\gamma_{zx}).$$

By using the constitutive laws (6.18) and (6.20) for an isotropic medium, we obtain for $W$ an expression depending only on the stresses:

$$W = \frac{1}{2E}\,\sigma_z^2 + \frac{1}{2G}\,(\tau_{xz}^2 + \tau_{yz}^2). \tag{6.77}$$

Since $\sigma_z \equiv 0$ in pure torsion, the first term does not give rise to any coupling. On the other hand, the shear stresses are

$$\tau_{xz} = E\,\frac{\partial\Phi}{\partial x} + G\,\frac{\partial K}{\partial y} + G\hat{\theta}\,\frac{\partial\Theta}{\partial y},$$

$$\tau_{yz} = E\,\frac{\partial\Phi}{\partial y} - G\,\frac{\partial K}{\partial x} - G\hat{\theta}\,\frac{\partial\Theta}{\partial x},$$

and we obtain the following terms for the energetic coupling between bending without torsion and pure torsion:

$$W' = E\hat{\theta}\,\left(\frac{\partial\Phi}{\partial x}\frac{\partial\Theta}{\partial y} - \frac{\partial\Phi}{\partial y}\frac{\partial\Theta}{\partial x}\right) + G\hat{\theta}\,\left(\frac{\partial K}{\partial x}\frac{\partial\Theta}{\partial x} + \frac{\partial K}{\partial y}\frac{\partial\Theta}{\partial y}\right).$$

Stokes' formula

$$\int_D \left(\frac{\partial\Phi}{\partial x}\frac{\partial\Theta}{\partial y} - \frac{\partial\Phi}{\partial y}\frac{\partial\Theta}{\partial x}\right) dxdy = \sum_0^n \int_{C_i} \Phi \, d\Theta = 0$$

shows that the contribution of the first term is zero when it is integrated over the cross section. Except for a factor $G\hat{\theta}$ the contribution of the second may be written

$$\int_D \text{grad } K \cdot \text{grad } \Theta \, dxdy = -\int_D K\nabla^2\Theta \, dxdy + \sum_0^n \int_{C_i} K \frac{\partial\Theta}{\partial n} ds.$$

In view of equations (6.49)', (6.49)", and (6.76) this vanishes if

$$\int_D K \, dxdy + \sum_1^n \beta_i\Omega_i = 0. \tag{6.78}$$

This is the condition found by Trefftz; it neither determines the parameter $c$ nor defines bending without torsion except implicitly.

On the other hand, if one writes (6.78) in the form

$$\int_D \text{grad } K \cdot \text{grad } \Theta \, dxdy = -\int_D \Theta\nabla^2 K \, dxdy + \sum_0^n \int_{C_i} \Theta \frac{\partial K}{\partial n} ds$$

and uses (6.76) and (6.76)', it becomes (after division by $2\nu$) the equivalent decoupling condition

$$\int_D \Theta \, (bx - ay + c)dxdy + \sum_1^n \alpha_i\Omega_i(b\hat{x}_i - a\hat{y}_i + c)$$

$$\equiv \int_{\Omega_o} \hat{\Theta} \, (bx - ay + c)dxdy = 0. \tag{6.79}$$

This supplies an explicit value of $c$ from its rearranged form

$$cJ = 2a \int_{\Omega_o} \hat{\Theta} \, y \, dxdy - 2b \int_{\Omega_o} \hat{\Theta} \, x \, dxdy, \tag{6.80}$$

and also a kinematic interpretation of bending without torsion. Indeed, it states that the weighted average of the twist of the fibers as given by (6.75) is annulled:

$$\int_{\Omega_0} \frac{\partial \omega_z}{\partial z} \hat{\Theta} \, dx \, dy = 0. \tag{6.81}$$

Since the weight function is the extended stress function, the average includes the twist of the virtual fibers located in the cavities. In Timoshenko's definition, on the other hand, we deal with the ordinary medium and only real fibers. Apart from its most important property, which is to render additive the energies of bending and torsion, the definition by Trefftz of bending without torsion has the following two characteristics which are interesting because of their simplicity:

i) <u>The static equivalence to zero of the secondary field.</u> Whatever the value of the parameter $c$, the shear forces of the secondary field are zero, because by Stokes' formula

$$G \int_D \frac{\partial K}{\partial y} \, dxdy = G \sum_0^n \int_{C_i} x \, dK = 0,$$

$$- G \int_D \frac{\partial K}{\partial x} \, dxdy = G \sum_0^n \int_{C_i} y \, dK = 0.$$

By contrast, the twisting moment of the secondary field does depend on the parameter:

$$-G \int_D \left(x \frac{\partial K}{\partial x} + y \frac{\partial K}{\partial y}\right) dxdy = G \int_D K \nabla^2 \left(\frac{x^2+y^2}{2}\right) dxdy$$

$$- G \sum_0^n \int_{C_i} K \, (xdy - ydx) = 2G\left\{\int_D K \, dxdy + \sum_1^n \beta_i \Omega_i\right\}.$$

However when Trefftz' condition (6.78) is satisfied, the expression on the right-hand side vanishes.

ii) <u>The identity of the center of bending and the center of torsion.</u> The center of bending $(x_F, y_F)$ is the point of the cross section through which pass the lines of action of the shear forces so as to form a system statically equivalent to the shear stress field. Because of the preceding result it is enough to examine the equivalence with the principal field. The identity of the moments with respect to the centroid is expressed by

$$E \int_D \left(x \frac{\partial \Phi}{\partial y} - y \frac{\partial \Phi}{\partial x}\right) dxdy = -y_F T_x + x_F T_y.$$

By transforming the left hand side by Stokes' formula, using (6.40) in the other side, and dropping the factor $E$, we obtain

$$-\sum_0^n \int_{c_i} \Phi(xdx + ydy) = a(-y_F I_{xx} + x_F I_{xy}) + b(-y_F I_{xy} + x_F I_{yy}).$$

It is therefore reasonable to decompose $\Phi$ into its parts with the parameters $a$ and $b$

$$\Phi = a\Phi_a + b\Phi_b$$

controlled, according to (6.48), by the equations

$$\nabla^2 \Phi_a = -x, \quad \partial\Phi_a/\partial n = 0 \quad \text{on} \quad c_i, \quad i = 0,1,\dots,n, \tag{6.82}$$

$$\nabla^2 \Phi_b = -y, \quad \partial\Phi_b/\partial n = 0 \quad \text{on} \quad c_i, \quad i = 0,1,\dots,n. \tag{6.83}$$

Identification of the coefficients of $a$ and $b$ yields the following equations for determining the center of bending

$$\begin{aligned} y_F I_{xx} - x_F I_{xy} &= \sum_0^n \int_{c_i} \Phi_a(x\,dx + y\,dy) \\ y_F I_{xy} - x_F I_{yy} &= \sum_0^n \int_{c_i} \Phi_b(x\,dx + y\,dy). \end{aligned} \tag{6.84}$$

In showing that the right hand sides are equivalent to those of equations (6.74) one proves the identity of the center of bending and the center of torsion of Weinstein-Kappus. To this end we multiply the first of equations (6.82) by the torsional warping function $H_\theta$ to obtain

$$H_\theta \nabla^2 \Phi_a = -xH_\theta$$

and integrate using Green's second formula to obtain

$$\int_D H_\theta x \, dxdy = -\int_D \Phi_a \nabla^2 H_\theta \, dxdy + \sum_0^n \int_{C_i} (\Phi_a \frac{\partial}{\partial n} H_\theta - H_\theta \frac{\partial}{\partial n} \Phi_a) ds.$$

Finally, in view of (6.67) and the second of equations (6.82) this becomes

$$\int_D H_\theta x \, dxdy = \sum_0^n \int_{C_i} \Phi_a (x \, dx + y \, dy).$$

In a similar way one establishes the equivalence of the other two right hand sides of (6.74) and (6.84).

The equivalencies in question were discovered by *Cicala*. Using Timoshenko's definition, he considers our $(x_F, y_F)$ as only the principal values of the coordinates of the center of bending; a secondary part depends on Poisson's ratio. This is another advantage of Trefftz's definition, which makes the center of bending and torsion independent of Poisson's effects.

The determination of the warping due to bending may be envisaged in several ways. If the potential of the principal shear stress field has been determined one has

$$g = 2(1+\nu)\Phi + H + \lambda,$$

and it is enough to recover the harmonic function $H$ with Neumann's data

$$\frac{\partial H}{\partial n} = \nu \frac{\partial}{\partial s}\{\frac{x^2+y^2}{2} c - \frac{a}{6}(y^3+3yx^2) + \frac{b}{6}(x^3+3xy^2)\} \quad \text{on } C_i, \quad i = 0,1,\dots,n,$$

which contributes a supplementary Poisson's effect.

If a sufficiently simple formal solution exists one may

prefer the transition to the conjugate harmonic function $Z$ with Dirichlet's data.

One may also seek at once the total warping by its Poisson's equation with Neumann's boundary data (6.64).

## 6.13. The Stiffness Relation for the Twist

We have seen that the parameters $(a,b)$ had a kinematic interpretation as derivatives of curvatures (6.38) and an interpretation in terms of shear forces connected by the stiffness relations (6.40).

We are now in a position to extend this point of view to the parameter $\theta$, already defined from a kinematic point of view by (6.39) as the twist (per unit length) of the centroidal fiber. Superposing the pure torsion and the bending without torsion gives $\theta = \hat{\theta} + \nu c$, where for (6.65) the part $\hat{\theta}$ due to pure torsion is connected with the resultant couple $M_F$ of the shear stresses about the center of bending-torsion $(x_F, y_F)$ as follows:

$$GJ\hat{\theta} = M_F = \int_D [(x-x_F)\tau_{yz} - (y-y_F)\tau_{xz}]dxdy. \qquad (6.85)$$

The parameter $c$ depends on $a$ and $b$ according to (6.80), which we rewrite in the form

$$c = ay_0 - bx_0, \qquad (6.86)$$

defining the point $(x_0, y_0)$ by the formulas

$$Jy_0 = 2\int_{\Omega_0} \hat{\theta} y \, dx\, dy, \quad Jx_0 = 2\int_{\Omega_0} \hat{\theta} x \, dx\, dy. \qquad (6.87)$$

We may now rewrite the twist as

$$\partial\omega_z/\partial z = \hat{\theta} + \nu(bx - ay + c)$$

and, with reference to equation (6.79) for bending without torsion, redefine $\hat{\theta}$ by the weighted average

$$J\hat{\theta} = 2\int_{\Omega_0} \hat{\theta} \frac{\partial \omega_z}{\partial z} \, dx \, dy. \qquad (6.87)'$$

We thus complete the stiffness relations by

$$M_F/GJ = \theta + \nu(bx_0 - ay_0), \qquad (6.88)$$

which finally allows the interpretation of the three parameters in terms of the resultants of the shear stresses about the center of bending-torsion $(x_F, y_F)$.

## 6.14. Total Energy as a Function of the Deformations of the Fibers

The strain energy density as a function of the strains is

$$W = \frac{1}{2} E\varepsilon_z^2 + \frac{1}{2}G(\gamma_{xz}^2 + \gamma_{yz}^2)$$

where from (6.51) the specific elongation is

$$\varepsilon_z = \varepsilon + xK_x(z) + yK_y(z)$$

and, by superposition of torsion and bending without torsion, the shear strains are

$$\gamma_{xz} = 2(1+\nu)\frac{\partial\Phi}{\partial x} + \frac{\partial K}{\partial y} + \hat{\theta}\,\frac{\partial\Theta}{\partial y}, \quad \gamma_{yz} = 2(1+\nu)\frac{\partial\Phi}{\partial y} - \frac{\partial K}{\partial x} - \hat{\theta}\,\frac{\partial\Theta}{\partial x}.$$

By integration over the cross section, it has already been verified that the contribution of $\Theta$ was uncoupled from that of $\Phi$ and could be from that of $K$ by a judicious choice of the parameter $c$. The contributions of $\Phi$ and $K$ are in fact themselves uncoupled as Stokes' formula shows:

$$\int_D \left(\frac{\partial\Phi}{\partial x}\frac{\partial K}{\partial y} - \frac{\partial\Phi}{\partial y}\frac{\partial K}{\partial x}\right)dxdy = \sum_0^n \int_{a_i} \Phi \, dK = 0.$$

The energy per unit length thus becomes

$$L = \int_D W\,dxdy = \tfrac{1}{2}E(\Omega\varepsilon^2 + I_{xx}K_x^2 + 2I_{xy}K_xK_y + I_{yy}K_y^2)$$
$$+ 2G(1+\nu)^2\int_D \text{grad}\Phi\cdot\text{grad}\Phi\,dxdy + \tfrac{1}{2}G\int_D \text{grad}K\cdot\text{grad}K\,dxdy$$
$$+ \frac{1}{2}G\hat{\theta}^2\int_D \text{grad}\,\Theta\cdot\text{grad}\,\Theta\,dxdy.$$

The interpretation of this formula in terms of deformations of the fibers appeals to the decomposition of $\Phi$ into its parts controlled by equations (6.82) and (6.83) and the analogous decomposition $K = aK_a + bK_b$ with

$$\nabla^2 K_a = y - y_o, \qquad \nabla^2 K_b = -x + x_o \qquad \text{in } D;$$
$$dK_a = 0, \qquad dK_b = 0, \qquad \text{on } c_i,\quad i = 0,1,\ldots,n;$$
$$\int_{c_i}\frac{\partial K_a}{\partial n}\,ds = \Omega_i(y_o - \hat{y}_i), \qquad \int_{c_i}\frac{\partial K_b}{\partial n}\,ds = \Omega_i(-x_o + \hat{x}_i),$$
$$i = 1,2,\ldots,n. \tag{6.89}$$

All the stress functions have thus received definitions of a purely geometric nature.

With reference to the result (6.72) and the new definitions

$$S_{xx} = 4(1+\nu)^2\int_D \text{grad}\Phi_a\cdot\text{grad}\Phi_a\,dxdy + \int_D \text{grad}\,K_a\cdot\text{grad}\,K_a\,dxdy,$$
$$S_{xy} = 4(1+\nu)^2\int_D \text{grad}\Phi_a\cdot\text{grad}\Phi_b\,dxdy + \int_D \text{grad}\,K_a\cdot\text{grad}\,K_b\,dxdy,$$
$$S_{yy} = 4(1+\nu)^2\int_D \text{grad}\Phi_b\cdot\text{grad}\Phi_b\,dxdy + \int_D \text{grad}\,K_b\cdot\text{grad}\,K_b\,dxdy, \tag{6.90}$$

several variants of which are accessible by Green's transformations, the energy per unit length becomes

$$L = \frac{E}{2}[\Omega\varepsilon^2 + I_{xx}(\ddot{u}_o)^2 + 2I_{xy}\ddot{u}_o\ddot{v}_o + I_{yy}(\ddot{v}_o)^2]$$
$$+ \frac{G}{2}[S_{xx}(\dddot{u}_o)^2 + 2S_{xy}\dddot{u}_o\dddot{v}_o + S_{yy}(\dddot{v}_o)^2 + J\hat{\theta}^2]. \tag{6.91}$$

The first term is the energy of extension with $\varepsilon$ as the specific elongation of the centoidal fiber; the following three represent an energy of bending generated by the curvatures of this fiber; we find next a group of three terms due to the variations of these curvatures and, finally, the term of the energy of torsion.

## 6.15. Total Energy as a Function of Generalized Forces

It is sometimes useful to replace deformations by generalized forces, such as the axial force, shear force, bending moment, and twisting moment, with the help of formulas such as (6.25) or (6.26) and (6.31)

For the twisting moment we use (6.85) with $M_F$ as the twisting moment with respect to the axis of bending-torsion. To keep the result simple we take the case in which the axes have been directed so as to annul the product of inertia $I_{xy}$. Then we have the *complementary* form for the energy per unit length

$$\Lambda = \frac{1}{2E\Omega} T_z^2 + \frac{1}{2EI_{xx}} M_x^2 + \frac{1}{2EI_{yy}} M_y^2$$
$$+ \frac{1}{2G}[S_{xx}(\frac{T_x}{I_{xx}})^2 + 2S_{xy} \frac{T_x}{I_{xx}} \frac{T_y}{I_{yy}} + S_{yy}(\frac{T_y}{I_{yy}})^2 + \frac{1}{J} M_F^2]. \tag{6.92}$$

## 6.16. The Generalized Constitutive Equations for Bending and Torsion of Beams

A development of the energy leading to a more fundamental interpretation starts from Clapeyron's theorem with the strains replaced by their expressions as derivatives of displacements; in the present case this yields

$$L_c = \frac{1}{2} \int_D [\sigma_z \frac{\partial w}{\partial z} + \tau_{xz}(\frac{\partial u}{\partial z} + \frac{\partial w}{\partial x}) + \tau_{yz}(\frac{\partial v}{\partial z} + \frac{\partial w}{\partial y})] dxdy.$$

Always keeping $I_{xy} = 0$ to simplify the exposition, we express the stresses as functions of the generalized forces:

$$\sigma_z = \frac{T_z}{\Omega} + \frac{M_x}{I_{xx}} x + \frac{M_y}{I_{yy}} y, \tag{6.93}$$

$$\tau_{xz} = \frac{T_x}{I_{xx}}\left[\frac{\partial \Phi_a}{\partial x} + \frac{1}{2(1+\nu)} \frac{\partial K_a}{\partial y}\right] + \frac{T_y}{I_{yy}}\left[\frac{\partial \Phi_b}{\partial x} + \frac{1}{2(1+\nu)} \frac{\partial K_b}{\partial y}\right] + \frac{M_F}{J} \frac{\partial \Theta}{\partial y},$$

$$\tau_{yz} = \frac{T_x}{I_{xx}}\left[\frac{\partial \Phi_a}{\partial y} - \frac{1}{2(1+\nu)} \frac{\partial K_a}{\partial x}\right] + \frac{T_y}{I_{yy}}\left[\frac{\partial \Phi_b}{\partial x} - \frac{1}{2(1+\nu)} \frac{\partial K_b}{\partial x}\right] - \frac{M_F}{J} \frac{\partial \Theta}{\partial x}, \tag{6.94}$$

and perform the integrations. A certain number of terms disappear by virtue of Stokes' formulas such as

$$\int_D \left(\frac{\partial \Theta}{\partial y} \frac{\partial w}{\partial x} - \frac{\partial \Theta}{\partial x} \frac{\partial w}{\partial y}\right) dxdy = \sum_0^n \int_{C_i} w \, d\Theta = 0,$$

and similar ones with $K_a$ or $K_b$ instead of $\Theta$. The non-vanishing terms give rise to the definitions of generalized displacements. For example, the development of the term in $\sigma_z \frac{\partial w}{\partial z}$ suggests the definitions

$$\Omega W(z) = \int_D w \, dxdy \tag{6.95}$$

for a mean longitudinal displacement $W(z)$, and

$$I_{xx}\ \alpha(z) = \int_D wx \, dxdy, \qquad I_{yy}\ \beta(z) = \int_D wy \, dxdy \tag{6.96}$$

for the mean rotations $\alpha(z)$ and $\beta(z)$ of the cross sections. These definitions reduce to identities when the section remains plane:

$$w = W + x\alpha + y\beta.$$

By using the equations (6.82) and (6.83) one obtains the equivalent definitions for the mean rotations

$$I_{xx}\alpha(z) = -\int_D w\nabla^2\Phi_a \, dxdy = \int_D \text{grad } w \cdot \text{grad } \Phi_a \, dxdy$$
$$I_{yy}\beta(z) = -\int_D w\nabla^2\Phi_b \, dxdy = \int_D \text{grad } w \cdot \text{grad } \Phi_b \, dxdy. \tag{6.96'}$$

The mean rotations occur here as weighted averages of the local rotation vector grad w, and are precisely those which allow the development of the terms $\tau_{xz}\ \partial w/\partial x$ and $\tau_{yz}\ \partial w/\partial y$.

Finally, the development of the terms in $\tau_{xz}\ \partial u/\partial z$ and $\tau_{yz}\ \partial v/\partial z$ suggests the introduction of the weighted means $U(z)$ and $V(z)$ of the transverse displacements:

$$I_{xx}U(z) = \int_D [(\frac{\partial\Phi_a}{\partial x} + \frac{1}{2(1+\nu)}\frac{\partial K_a}{\partial y})u + (\frac{\partial\Phi_a}{\partial y} - \frac{1}{2(1+\nu)}\frac{\partial K_a}{\partial x})v]dxdy,$$
$$I_{yy}V(z) = \int_D [(\frac{\partial\Phi_b}{\partial x} + \frac{1}{2(1+\nu)}\frac{\partial K_b}{\partial y})u + (\frac{\partial\Phi_b}{\partial y} - \frac{1}{2(1+\nu)}\frac{\partial K_b}{\partial x})v]dxdy. \tag{6.97}$$

The properties

$$\int_D \frac{\partial\Phi_a}{\partial x} dxdy = I_{xx}, \quad \int_D \frac{\partial\Phi_a}{\partial y} dxdy = \int_D \frac{\partial\Phi_b}{\partial x} dxdy = I_{xy},$$
$$\int_D \frac{\partial\Phi_b}{\partial y} dxdy = I_{yy}, \tag{6.98}$$

result naturally from the application of Green's first formula to the pairs of functions $(x,\Phi_a)$, $(y,\Phi_a)$, $(x,\Phi_b)$, and $(y,\Phi_b)$. Similarly, the properties

$$\int_D \frac{\partial K_a}{\partial x} dxdy = 0, \quad \int_D \frac{\partial K_a}{\partial y} dxdy = 0, \quad \int_D \frac{\partial K_b}{\partial x} dxdy = 0,$$
$$\int_D \frac{\partial K_b}{\partial y} dxdy = 0, \tag{6.98'}$$

result from Stokes' formula applied to the analogous pairs $(x,K_a)$, $(y,K_a)$, $(x,K_b)$, and $(y,K_b)$, and are also contained implicitly in the equivalence to zero of the secondary shear stress field. As a result the definitions (6.97) reduce to identities if $u$ and $v$ are independent of $x$ and $y$.

It remains to interpret the generalized displacement $\omega(z)$ associated with the twisting moment $M_F$:

$$J\omega(z) = \int_D \left(\frac{\partial\Theta}{\partial y}u - \frac{\partial\Theta}{\partial x}v\right)dxdy = \int_D \Theta\left(\frac{\partial v}{\partial x} - \frac{\partial u}{\partial y}\right)dxdy$$

$$+ \sum_0^n \int_{C_i} \Theta\left(u\,\frac{\partial y}{\partial n} - v\,\frac{\partial x}{\partial n}\right)ds$$

$$= 2\int_D \Theta\omega_z\; dxdy - \sum_1^n \alpha_i \int_{C_i} u\,dx + v\,dy = 2\int_{\Omega_0} \hat{\Theta}\omega_z\; dxdy. \qquad (6.98)$$

This displacement uses the same weighting for the rotation of the fibers as that already used to weight their twist. Once again this definition reduces to an identity if $\omega_z$ is independent of $x$ and $y$.

The energy per unit length now takes the form

$$L_c = \frac{1}{2}\left[T_z\,\frac{dW}{dz} + M_x\,\frac{d\alpha}{dz} + M_y\,\frac{d\beta}{dz} + T_x\left(\alpha + \frac{dU}{dz}\right) + T_y\left(\beta + \frac{dV}{dz}\right) + M_F\,\frac{d\omega}{dz}\right]. \qquad (6.99)$$

This is a canonical form involving a sum of products of *conjugate* variables. We may compare it with the complementary form (6.92). The latter being quadratic and homogeneous in the generalized forces, by Euler's theorem it can assume the form

$$\Lambda = \frac{1}{2}\left[T_z\,\frac{\partial\Lambda}{\partial T_z} + M_x\,\frac{\partial\Lambda}{\partial M_x} + M_y\,\frac{\partial\Lambda}{\partial M_y} + T_x\,\frac{\partial\Lambda}{\partial T_x} + T_y\,\frac{\partial\Lambda}{\partial T_y} + M_F\,\frac{\partial\Lambda}{\partial M_F}\right].$$

By identification with (6.99), this yields the generalized constitutive equations of Saint-Venant's theory

$$\frac{dW}{dz} = \frac{\partial\Lambda}{\partial T_z} = \frac{1}{E\Omega}\,T_z, \qquad (6.100)$$

$$\frac{d\alpha}{dz} = \frac{\partial\Lambda}{\partial M_x} = \frac{1}{EI_{xx}}\,M_x, \quad \frac{d\beta}{dz} = \frac{\partial\Lambda}{\partial M_y} = \frac{1}{EI_{yy}}\,M_y, \qquad (6.100)'$$

$$\alpha + \frac{dU}{dz} = \frac{\partial \Lambda}{\partial T_x} = \frac{S_{xx}}{G} \frac{T_x}{I_{xx}^2} + \frac{S_{xy}}{G} \frac{T_y}{I_{yy}^2} ,$$

$$\beta + \frac{dV}{dz} = \frac{\partial \Lambda}{\partial T_y} = \frac{S_{xy}}{G} \frac{T_x}{I_{xx}^2} + \frac{S_{yy}}{G} \frac{T_y}{I_{yy}^2} , \qquad (6.100)''$$

$$\frac{d\omega}{dz} = \frac{\partial \Lambda}{\partial M_F} = \frac{1}{GJ} M_F . \qquad (6.100)'''$$

The left-hand sides are the generalized deformations conjugate to the corresponding generalized forces. In particular, $\alpha + dU/dz$ and $\beta + dV/dz$ are the generalized deformations due to the shear forces; by construction, they are weighted means of shear strains:

$$\begin{aligned} I_{xx}(\alpha + \frac{dU}{dz}) &= \int_D [(\frac{\partial \Phi_a}{\partial x} + \frac{1}{2(1+\nu)} \frac{\partial K_a}{\partial y})\gamma_{xz} \\ &\quad + (\frac{\partial \Phi_a}{\partial y} - \frac{1}{2(1+\nu)} \frac{\partial K_a}{\partial x})\gamma_{yz}] dxdy \\ I_{yy}(\beta + \frac{dV}{dz}) &= \int_D [(\frac{\partial \Phi_b}{\partial x} + \frac{1}{2(1+\nu)} \frac{\partial K_b}{\partial y})\gamma_{xz} \\ &\quad + (\frac{\partial \Phi_b}{\partial y} - \frac{1}{2(1+\nu)} \frac{\partial K_b}{\partial x})\gamma_{yz}] dxdy . \end{aligned} \qquad (6.101)$$

It is interesing to verify that the integrals in (6.101) vanish when the shear strains are produced only by a state of pure torsion; by using equations (6.89) one can reduce them to expressions proportional to

$$\int_{\Omega_o} \hat{\Theta}(y-y_o)dxdy \qquad \text{and} \qquad \int_{\Omega_o} \hat{\Theta}(x-x_o)dxdy$$

which clearly vanish by virtue of (6.87).

## 6.17. One-Dimensional Formulation of Bending and Torsion of Beams

In terms of the mean displacements $U(z)$, $V(z)$, $W(z)$ and the mean rotations $\alpha(z)$, $\beta(z)$, $\omega(z)$, the constitutive equations and the equilibrium equations take the form of ordinary differential equations in the variable $z$. This presentation of Saint-Venant's theory should not obscure the fact that it is not an exact solution of the equations of elasticity unless the shear forces and the twisting moment are constant. Moreover, the stresses in the terminal sections should vary as prescribed by equation (6.23) for the axial stress and by the solutions of the problems governing the distribution of the shear stresses. Also, each section should be free to warp and to contract in its own plane.

These limitations are severe for those technical applications which entail non-vanishing tractions on the lateral surfaces and kinematic constraints or less restrictive statics at the terminal sections. It is possible to give Saint-Venant's theory the extensions necessary to allow variable shear forces and twisting moment while maintaining an exact solution. As Michell has shown, such an extension brings into play a connection between the problems of the harmonic functions of torsion and bending, as encountered in this chapter, and the problems of the biharmonic functions of states of plane stress. It should therefore preferably be started after a study of the latter. In spite of its complexity, this more general theory still does not allow a relaxation of the terminal constraints. It is customary here to appeal to *Saint-Venant's principle,* according to which every perturbation of the traction distribution on a terminal section, being necessarily realized by the addition of a distribution statically equivalent to zero, tends to disappear at a distance from the section of the same order of magnitude as the transverse dimensions.

An approximate theory results by combining the generalized constitutive equations (6.100) with the more general equations of equilibrium

$$dM_x/dz = T_x, \quad dM_y/dz = T_y, \quad dT_z/dz = 0,$$
$$dT_x/dz + p_x = 0, \quad dT_y/dz + p_y = 0, \quad dM_F/dz + m = 0$$

which contain the distributed loads $p_x(z)$, $p_y(z)$, and the distributed twisting moment $m(z)$. The justification of the theory relies on applications of the variational principles of the theory of elasticity and accordingly will not be discussed at this time. Such a one-dimensional formulation has the advantage of simple boundary conditions. A clamped end at $z = 0$, for example, is represented by the boundary conditions

$$U(0) = 0,\ V(0) = 0,\ W(0) = 0,\ \alpha(0) = 0,\ \beta(0) = 0,\ \omega(0) = 0.$$

The strain energy per unit length has already been presented in its complementary form (6.92) and in Clapeyron's form (6.99). It also clearly has a form directly expressed in terms of the generalized deformations

$$L = \frac{1}{2}E\Omega\left(\frac{dW}{dz}\right)^2 + \frac{1}{2}EI_{xx}\left(\frac{d\alpha}{dz}\right)^2 + \frac{1}{2}EI_{yy}\left(\frac{d\beta}{dz}\right)^2 + \frac{1}{2}GJ\left(\frac{d\omega}{dz}\right)^2$$
$$+ \frac{G}{2(S_{xx}S_{yy}-S_{xy}^2)}\left[I_{xx}^2 S_{yy}\left(\alpha + \frac{dU}{dz}\right)^2 + I_{yy}^2 S_{xx}\left(\beta + \frac{dV}{dz}\right)^2\right.$$
$$\left. - (I_{xx}^2 + I_{yy}^2)S_{xy}\left(\alpha + \frac{dU}{dz}\right)\left(\beta + \frac{dV}{dz}\right)\right]. \quad (6.101)$$

In certain conditions, which we will analyze, the strain energy can be simplified further by suppressing the part relating to deformations caused by shear forces. Let us study the case of a beam which is clamped at the section $z = 0$

and subjected by a shear force $T_x$ at the other end $z = \ell$. By integration of the first of the equilibrium equations (6.28) we find

$$M_x = T_x(z - \ell)$$

and then, by integration of the deformations,

$$\alpha(z) = \frac{T_x}{EI_{xx}}\left(\frac{1}{2}z^2 - \ell z\right), \qquad U(\ell) = \frac{T_x \ell^3}{3EI_{xx}} + \frac{S_{xx}T_x\ell}{GI_{xx}^2}.$$

The beam will be called *long* if the second term in $U(\ell)$ is small compared with the first, that is to say, if the transverse displacement is due mainly to the curvature of the fibers and not to the deformation caused by shear forces. This will hold if in turn

$$\frac{6S_{xx}(1+\nu)}{I_{xx}\ell^2} << 1 \quad \text{or} \quad \ell >> \left[\frac{6S_{xx}(1+\nu)}{I_{xx}}\right]^{1/2}.$$

When a beam is long with respect to transverse dimensions we may thus neglect the deformations caused by the shear forces and write $\alpha = -dU/dz$, $\beta = -dV/dz$, which amounts to expressing an average orthogonality of the cross sections with respect to the fibers. The simplified energy is then a function of the mean curvatures:

$$L = \frac{1}{2}E\Omega\left(\frac{dW}{dz}\right)^2 + \frac{1}{2}EI_{xx}\left(\frac{d^2U}{dz^2}\right)^2 + \frac{1}{2}EI_{yy}\left(\frac{d^2V}{dz^2}\right)^2 + \frac{1}{2}GJ\left(\frac{d\omega}{dz}\right)^2. \quad (6.101)'$$

## 6.18. Applications

As we have already seen, practical determination of the stress functions for torsion and bending involves the solution of elliptic problems of Poisson or Laplace type with boundary data of Dirichlet or Neumann type. The current preference is more and more for a purely numerical solution, starting with either a finite difference scheme or a collection of finite elements. The variational principles controlling the functions being sought are therefore very important and may be presented as special cases of the general variational principles of linear elasticity.

It is still necessary to mention the existence of numerous analogies with other physical phenomena which may be understood experimentally, such as the deflection of a membrane or of a free or pressurized fluid interface, or the electric potential field in a dielectric enclosure. On this subject it is helpful to consult the chapter on analogies by *Mindlin* and *Salvadori* in the book by *Hetenyi*.

In the following examples, the method used will be strictly inverse: one starts with simple analytic solutions of Laplace's equation and studies the geometric forms which allow them to solve the proposed problem.

A. Stress function for torsion of the elliptic bar. Here we start with the fact that the warping function $H_\theta(x,y)$ is the harmonic conjugate of $Z_\theta(x,y) = \Theta(x,y) + \frac{1}{2}(x^2+y^2)$. Simple solutions are then furnished by polynomials in the complex variable $x + iy$, of which a typical term will be

$$H_\theta + iZ_\theta = (p_m + iq_m)(x + iy)^m$$

with

$$H_\theta = p_m\left(x^m - \frac{m(m-1)}{2}x^{m-2}y^2 \ldots\right) - q_m(mx^{m-1}y + \ldots)$$

$$Z_\theta = p_m(mx^{m-1}y + \ldots) + q_m\left(x^m - \frac{m(m-1)}{2}x^{m-2}y^2 \ldots\right).$$

Collecting for example the contributions of $q_0$ and $q_2$, the stress function

$$\Theta = q_0 + q_2(x^2 - y^2) - \frac{1}{2}(x^2 + y^2)$$

takes the value zero on the ellipse $x^2/a^2 + y^2/b^2 = 1$ provided we choose

$$q_0 = \frac{a^2b^2}{a^2+b^2}, \quad 2q_2 = \frac{a^2-b^2}{a^2+b^2} .$$

By its construction, the function

$$\Theta = \frac{a^2b^2}{a^2+b^2}\left(1 - \frac{x^2}{a^2} - \frac{y^2}{b^2}\right)$$

is regular in the interior of the ellipse, vanishes on its boundary, and automatically satisfies (6.49)'; it is therefore Prandtl's stress function for the torsion of a bar with elliptic section. The associated warping function is, up to a constant,

$$H_\theta = - \frac{a^2 - b^2}{a^2 + b^2} xy.$$

The shear stresses follow by partial differentiation of either $\Theta$ or $H_\theta$:

$$\frac{\tau_{xz}}{G\theta} = \frac{\partial\Theta}{\partial y} = \frac{\partial H_\theta}{\partial x} - y = - \frac{2a^2y}{a^2+b^2} ,$$

$$\frac{\tau_{yz}}{G\theta} = - \frac{\partial\Theta}{\partial n} = \frac{\partial H_\theta}{\partial y} + x = \frac{2b^2x}{a^2+b^2} .$$

With $a > b$ the stress with greatest absolute value is $\tau_{xz}$ at $y = \pm b$.

The torsional stiffness constant requires the evaluation of the double integral of Prandtl's function over the section. By putting

$$x = \rho a \cos\theta, \quad y = \rho b \sin\theta, \quad \frac{D(x,y)}{D(\rho,\theta)} = ab\rho,$$

we find

$$J = 2\int_0^2 d\theta \int_0^1 \Theta\, ab\rho\, d\rho = \frac{2a^3b^3}{a^2+b^2} 2\pi \int_0^1 (1-\rho^2)\rho d\rho = \frac{\pi a^3b^3}{a^2+b^2} .$$

By symmetry, the center of torsion is located at the origin.

The warping function is single-valued and regular in the entire interior of the ellipse. In order to transform the elliptic bar into an elliptic tube, therefore, it suffices to take away the material contained in the curve with equation $\Theta = \alpha$, a positive constant to be chosen. This curve is another ellipse $x^2/m^2 + y^2/n^2 = 1$ with

$$\frac{m}{a} = \frac{n}{b} = \mu < 1, \quad \alpha = (1-\mu^2)\frac{a^2b^2}{a^2+b^2} .$$

The extended Prandtl's function $\hat{\Theta}$ is accordingly one with the value

$$\frac{a^2b^2}{a^2+b^2}(1 - \rho^2)$$

in the region $\mu \leq \rho \leq 1$, and $\mu$ in the cavity. For the torsional stiffness of the elliptic tube we easily find

$$J = 2\int_{\Omega_0} \hat{\Theta}\, dxdy = \frac{\pi a^3b^3}{a^2+b^2}(1 - \mu^4).$$

This result shows clearly that a significant portion of the central region may be removed without appreciably decreasing the torsional stiffness.

B. Stress functions for torsion of the circular bar and tube. It is enough to set $b = a$ in the preceding results. The level curves of resultant shear stress are circles centered at the origin. The most notable occurrence is the disappearance of warping. As shown earlier, the circular bar and tube are the only prismatic objects with a state of torsion compatible with rigid terminal constraint.

C. Stress functions with poles. In the expression for $H_\theta + iZ_\theta$ it is possible to have negative powers of $(x+iy)$

or, more generally, of $(x-x_0) + i(y-y_0)$, provided that the pole at $x = x_0$, $y = y_0$ falls outside the elastic region. The pole may be located outside the directrix $c_0$ or inside a cavity. The typical contribution of a simple pole at the origin is

$$H_\theta + iZ_\theta = \frac{p_{-1} + iq_{-1}}{(x+iy)} = \frac{(p_{-1}+iq_{-1})(x-iy)}{x^2+y^2}.$$

Consider for example the case of a stress function carrying the contribution of a simple pole at $x = a$, $y = 0$ and another at $x = b$, $y = 0$:

$$\Theta = \frac{p(x-a)}{(x-a)^2+y^2} + \frac{q(x-b)}{(x-b)^2+y^2} - \frac{1}{2}(x^2+y^2) + hx + k.$$

We normalize the disposable parameters $a,b,p,q,h$ and $k$ by requiring that $\Theta$ vanish on a circle $x^2+y^2 = r^2$. With $y^2$ replaced by $r^2-x^2$, the function

$$\frac{p(x-a)}{r^2+a^2-2ax} + \frac{q(x-b)}{r^2+b^2-2bx} - \frac{1}{2}r^2 + hx + k$$

should vanish identically. This implies notably the condition $abh = 0$, resulting from the annihilation of the term in $x^3$ in the numerator after reducing to a common denominator. There is no real distinction between the possibilities $a = 0$ or $b = 0$ (since either would place one pole at the origin) but the alternative $h = 0$ must separately be taken under consideration.

If $b = 0$, one infers from the vanishing of coefficients of the $x^2$, $x$, and constant terms, that

$$p = a(2k-r^2), \quad q = -hr^2, \quad (2k-r^2)(a^2-r^2) = 0.$$

Suppose first that $k = r^2/2$, which implies that $p = 0$.

Then there is only the pole at the origin. The stress function

$$\Theta = \frac{-hr^2x}{x^2+y^2} - \frac{1}{2}(x^2+y^2) + hx + \frac{1}{2}r^2$$

vanishes not only on the circle $x^2+y^2 = r^2$, but also on the circle $(x-h)^2 + y^2 = h^2$, and represents that of a shaft of radius $h$ carrying a circular indentation of radius $r$ centered on its circumference. (The cross section is similar to that of a shaft with key.) Up to a constant, the warping function is

$$H = (1 - \frac{r^2}{x^2+y^2})hy.$$

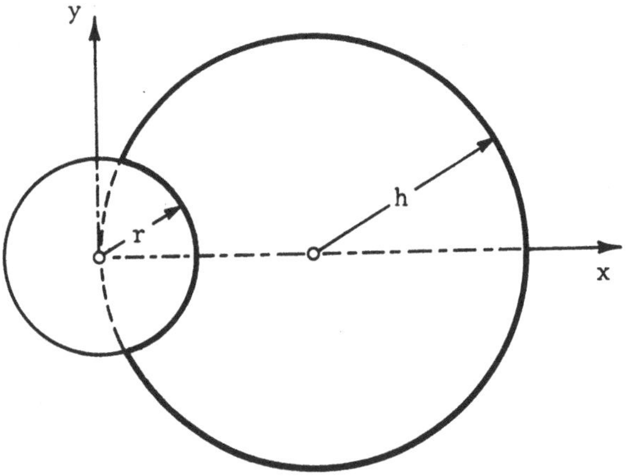


Figure 6.6

Second, suppose that $a = r$. The stress function becomes

$$\Theta = \frac{r(2k-r^2)(x-r)}{(x-r)^2 + y^2} - \frac{hr^2x}{x^2+y^2} - \frac{1}{2}(x^2+y^2) + hx + k$$

$$= (x^2+y^2-r^2)\left\{\frac{k-r^2+xr-\frac{1}{2}(x^2+y^2)}{(x-r)^2 + y^2} + \frac{hx}{x^2+y^2}\right\}.$$

This vanishes not only on $x^2+y^2=r^2$, but also on the curve with equation

$$-\frac{1}{2}+\frac{k-r^2/2}{(x-r)^2+y^2}+\frac{hx}{x^2+y^2}=0.$$

For $h = 0$ one loses the pole at the origin, and the latter curve becomes

$$(x-r)^2+y^2=2k-r^2,$$

which is a circle centered at the remaining pole. It forms a key notch in a shaft of radius $r$ centered at the origin and the solution is thus not essentially new.

On the other hand, by choosing

$$k=r^2-\frac{1}{2}hr$$

one loses no pole and the function again vanishes on the two circles $x^2+y^2-2rx=0$ (centered at the pole $x = r$ and passing through the pole at the origin), and $(x-h)^2+y^2 = h(h-r)$ (centered at $x = h$).

As regions excluding the poles one may find a circle with key notch centered at an interior point,

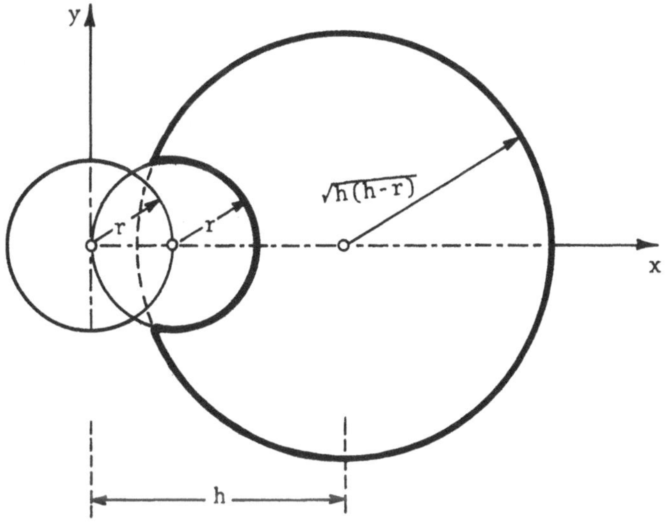


Figure 6.7

a curvilinear triangle,

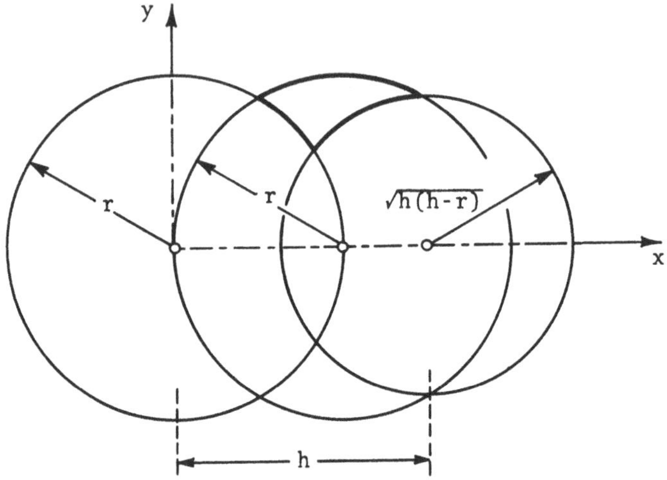


Figure 6.8

or a portion of an eccentric annulus.

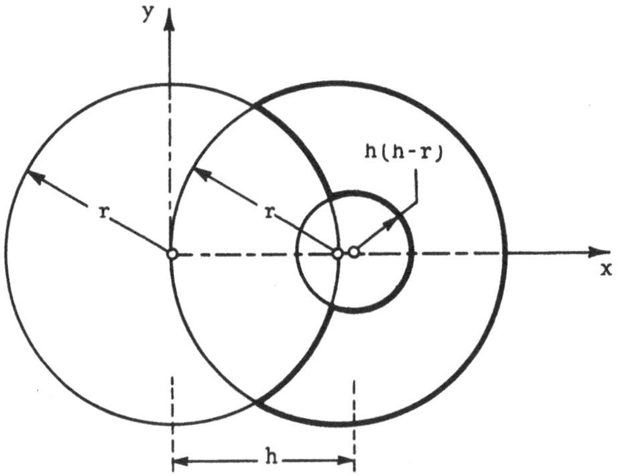


Figure 6.9

Now instead of $b = 0$ we study now the situation when $h = 0$. The remaining conditions that the function

$$\Theta = \frac{p(x-a)}{(x-a)^2+y^2} + \frac{q(x-b)}{(x-b)^2+y^2} - \frac{1}{2}(x^2+y^2) + k$$

should vanish on the circle $x^2+y^2 = r^2$ reduce to a system of homogeneous linear equations in the unknowns $p$, $q$, and $\lambda = k - r^2/2$:

$$-2bp - 2aq + 4ab\lambda = 0,$$

$$(r^2+b^2+2ab)p + (r^2+a^2+2ab)q - [2b(r^2+a^2)+2a(r^2+b^2)]\lambda = 0,$$

$$-a(r^2+b^2)p - b(r^2+a^2)q + (r^2+a^2)(r^2+b^2)\lambda = 0.$$

Evaluation of the determinant yields

$$\Delta = 2(a-b)(a^2-r^2)(b^2-r^2)(r^2-ab).$$

In order to avoid the preceding cases and to find poles not located on the base circle, we make the equations compatible by choosing $r^2 = ab$. Then one finds

$$p = a\lambda, \quad q = b\lambda, \quad k = \frac{1}{2} r^2 + \lambda.$$

It remains to be seen on which other loci the function

$$\Theta = \lambda[1 + \frac{a(x-a)}{(x-a)^2+y^2} + \frac{b(x-b)}{(x-b)^2+y^2}] + \frac{1}{2}(r^2 - x^2 - y^2)$$

vanishes or takes a constant value, say $\alpha$. This amounts to studying the equation

$$(x^2+y^2-r^2)\left\{\frac{\lambda(x^2+y^2)+ab-(a+b)x)}{[(x-a)^2+y^2][(x-b)^2+y^2]} - \frac{1}{2}\right\} = \alpha.$$

One would expect, for example, that this equation would hold approximately on a small circle $(x-a)^2 + y^2 = \varepsilon^2$ excluding the pole so as to obtain the stress function for a shaft pierced by a small, off-center hole. Substituting for $y^2$ yields the expressions

$$x^2 + y^2 - r^2 = \varepsilon^2 - a^2 - r^2 + 2ax,$$

$$x^2 + y^2 + ab - (a+b)x = \varepsilon^2 - a^2 + 2ax + ab - (a+b)x,$$

$$(x-b)^2 + y^2 = \varepsilon^2 - a^2 + 2ax + b^2 - 2bx.$$

If $\varepsilon$ is small compared with $(b-a)$ one may neglect the terms in $\varepsilon^2$ and the above respectively reduce to

$$-a(a+b-2x), \quad (b-a)(a-x), \quad (b-a)(a+b-2x).$$

It is thus necessary to satisfy the equation

$$-a(a+b-2x)\left[\frac{\lambda(a-x)}{\varepsilon^2(a+b-2x)} - \frac{1}{2}\right] = -\frac{a\lambda}{\varepsilon^2}(a-x) + \frac{a}{2}(a+b-2x) = \alpha$$

which implies that $\lambda = \varepsilon^2$ and $\alpha = a(b-a)/a$.

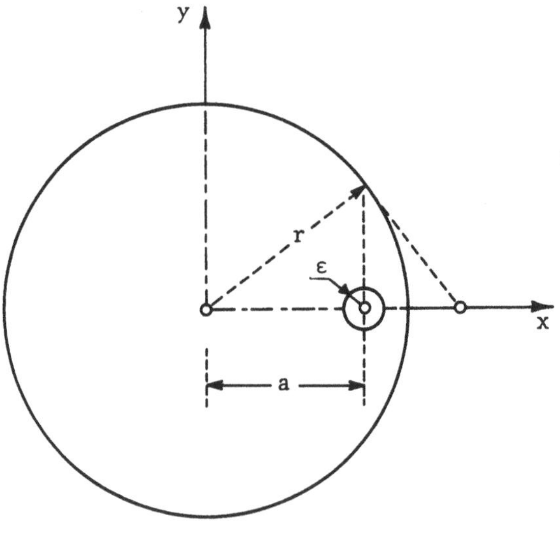


Figure 6.10

At the point $x = a + \varepsilon,\ y = 0$ the shear stress, neglecting terms in $\varepsilon^2$, is

$$\tau_{yz} = G\hat{\theta}(2a + \varepsilon).$$

By comparison, without the hole, the value there is

$$\tau_{yz} = G\hat{\theta}(a + \varepsilon).$$

D. Torsion of a triangular bar.

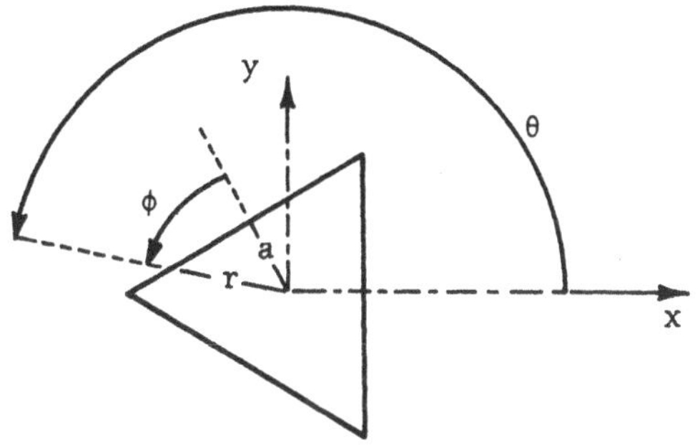


Figure 6.11

Let

$$r \cos \phi = r \cos(\theta - \theta_i) = x \cos \theta_i + y \sin \theta_i = a$$

be the equation of the $i^{th}$ side of a triangle, situated at a distance $a$ from the origin with its normal making an angle $\theta_i$ with the $x$-axis. If the origin is chosen to be equidistant from the three sides, then the function

$$A \prod_{i=1}^{3} (x \cos \theta_i + y \sin \theta_i - a)$$

vanishes on the boundary; we study under what conditions it could become the torsion function of that bar. Without loss of generality $\theta_1 = 0$ and the expanded product is

$$\begin{aligned} A[&x^3\cos\theta_2\cos\theta_3 + x^2y\sin(\theta_2+\theta_3) + xy^2\sin\theta_2\sin\theta_3 \\ &- ax^2(\cos\theta_2+\cos\theta_3+\cos\theta_2\cos\theta_3) - ay^2\sin\theta_2\sin\theta_3 \\ &- axy(\sin\theta_2+\sin\theta_3+\sin(\theta_2+\theta_3)) + a^2y(\sin\theta_2+\sin\theta_3) \\ &+ a^2x(1+\cos\theta_2+\cos\theta_3) - a^3]. \end{aligned}$$

In order to be a stress function this expression should be of the form $Z_\theta(x,y) - (x^2+y^2)/2$, where $Z_\theta(x,y)$ is harmonic. It is thus necessary to recover the harmonic combinations

$x^3 - 3xy^2$ and $y^3 - 3x^2y$, which require, respectively,

$$\sin\theta_2 \sin\theta_3 = -3\cos\theta_2 \cos\theta_3 \quad \text{or} \quad \tan\theta_2 \tan\theta_3 = -3,$$
$$\sin(\theta_1 + \theta_2) = 0,$$

and impose the angles $\theta_2 = 120^\circ$ and $\theta_3 = 240^\circ$. The section must therefore be an equilateral triangle. Then we have

$$\frac{1}{4} A(x^3 - 3xy^2) + \frac{3}{4} Aa(x^2 + y^2) - Aa^3.$$

The desired form also requires $A = -2/3a$.

Definitively, then, the stress function is

$$\Theta = -\frac{1}{6a}(x^3 - 3xy^2) - \frac{1}{2}(x^2 + y^2) + \frac{2}{3} a^2.$$

Since

$$(x+iy)^3 = x^3 - 3xy^2 + i(3xy^2 - y^3)$$

the normalized warping function is, up to a constant,

$$H = \frac{a}{6}(3xy^2 - y^3).$$

The torsional stiffness constant is $J = 9\,\frac{\sqrt{3}}{5}\,a^4$.

E. Torsion of a rectangular bar. In this case the boundaries of the section are chosen as $x = \pm a$ and $y = \pm b$. In order to treat this problem by development in a series of characteristic functions we endeavor to formulate it in such a way that the partial differential equation is homogeneous with separable variables and a pair of boundary conditions is homogeneous. For example, if $\Theta = M + a^2 - x^2$ then the function $M$ should satisfy Laplace's equation $\nabla^2 M = 0$ and should vanish for $x = a$. On the other hand, it should assume the values $M = x^2 - a^2$ at $y = \pm b$.

By separation of variables, the solution of type $M = f(x)g(y)$ of Laplace's equation reduces to that of the differential equations

$$-\frac{1}{f}\frac{d^2f}{dx^2} = \frac{1}{g}\frac{d^2g}{dy^2} = \alpha^2$$

where $\alpha$ is a constant. The problem being invariant under the interchanges of $x$ and $-x$ and of $y$ and $-y$, the solution should have the same properties; therefore the solutions of the differential equations to be kept are respectively $\cos \alpha x$ and $\cosh \alpha y$.

The choice of the sign of the separation constant $\alpha^2$ is now justified by the fact that it allows meeting the boundary conditions at $x = \pm a$ by choosing $\alpha a = n\pi/2$, $n = 1,3,5,\ldots$ . The solution of the problem thus takes the form

$$M = \sum_{1,3\ldots} A_n \cos \frac{n\pi}{2a} x \cosh \frac{n\pi}{2a} y.$$

The coefficients $A_n$ are to be determined by the boundary conditions for $y = \pm b$:

$$x^2 - a^2 = \sum_{1,3\ldots} (A_n \cosh \frac{n\pi b}{2a}) \cos \frac{n\pi x}{2a} .$$

The cosine functions being orthogonal in the interval $(-\pi/2, \pi/2)$, the coefficients may be found by Fourier's method of multiplying by $\cos(m\pi x/2a)$ and integrating from $-a$ to $a$. The result is

$$A_n \cosh \frac{n\pi b}{2a} = (-1)^{\frac{n+1}{2}} \frac{32a^2}{(n\pi)^3},$$

whence the torsion function becomes

$$\Theta = a^2 - x^2 + \sum_{1,3\ldots} (-1)^{\frac{n+1}{2}} \frac{32a^2}{(n\pi)^3} \frac{\cosh \frac{n\pi}{2a} y}{\cosh \frac{n\pi b}{2a}} \cos \frac{n\pi}{2a} x.$$

By interchanging in this formula the roles of $x$ and $y$, and a and b, one obtains a similar expansion.

By adding $\frac{1}{2}(x^2+y^2)$ to this formula, we obtain the harmonic conjugate of the warping function. The warping function itself then reads

$$H_\theta = xy + \sum_{1,3\ldots} (-1)^{\frac{n+1}{2}} \frac{32a^2}{(n\pi)^3} \frac{\sinh \frac{n\pi}{2a} y}{\cosh \frac{n\pi b}{2a}} \sin \frac{n\pi}{2a} x.$$

The shear stress is greatest at the midpoint of a long side. Thus, if $b > a$, with $x = a$ and $y = 0$,

$$\left(\frac{\tau_{yz}}{G\theta}\right)_{max} = 2a - \sum_{1,3\ldots} \frac{16a}{(n\pi)^2} \frac{1}{\cosh \frac{n\pi b}{a}} .$$

The torsional stiffness constant is

$$J = \frac{8}{3} ba^3 - \frac{1024}{\pi^5} \sum_{1,3\ldots} \frac{1}{n^5} \tanh \frac{n\pi b}{2a} .$$

F. Bending of a circular bar. By symmetry, a

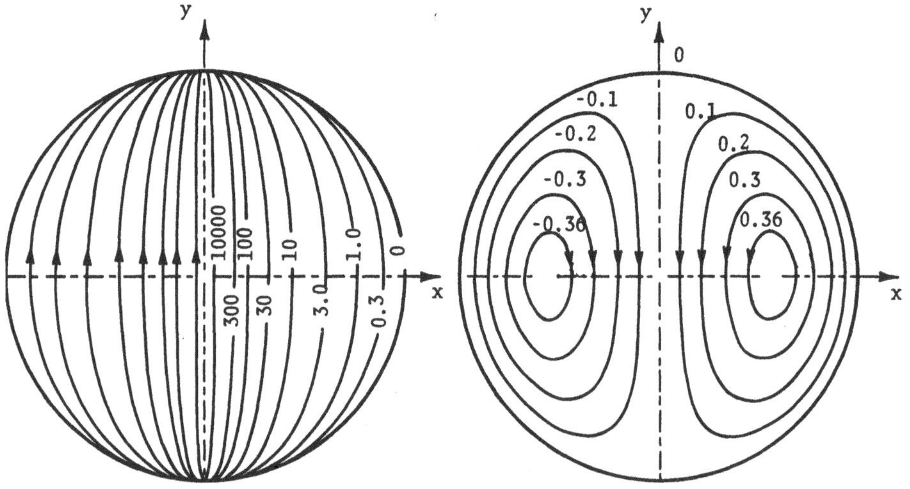


PRINCIPAL FIELD

STATE OF DOUBLE TORSION OF THE SECONDARY FIELD

Figure 6.12

circular bar has $I_{xy} = 0$ and $c = 0$ whatever the orientation of the axes issuing from the origin of the circle of radius $R$. It is thus sufficient to consider the case of a shear force along $Oy$, when $a = 0$ and $b = T_y/I_{yy}$. The bending function of the principal field solves the problem

$$\nabla^2\Phi = -by \quad \text{with} \quad \partial\Phi/\partial r = 0 \quad \text{for} \quad r = R.$$

With $\Phi = -by^3/6 + M$ we have

$$\nabla^2 M = 0 \quad \text{with} \quad \frac{\partial M}{\partial r} = \frac{1}{2}\, by^2 \sin\theta = \frac{1}{2}\, bR^2 \sin^3\theta \quad \text{for} \quad r = R.$$

Since $4 \sin^3\theta = 3 \sin\theta - \sin 3\theta$, we have an indication of the powers of $x + iy = re^{i\theta}$ to be chosen for the harmonic function $M$. With $p$ and $q$ real, we use

$$M = \operatorname{Im}(pre^{i\theta} + qr^3 e^{3i\theta}) = pr \sin\theta + qr^3 \sin 3\theta,$$

$$\frac{\partial M}{\partial r} = p \sin\theta + 3qr^2 \sin 3\theta.$$

By comparison at the limit $r = R$, $p = 3bR^2/8$ and $q = -b/24$. After reduction the bending function of the principal field is thus

$$\Phi = \frac{1}{8}\, b(3R^2 - r^2) r \sin\theta = \frac{1}{8}\, b(3R^2 - x^2 - y^2) y.$$

The trajectories of the stresses, the curves everywhere tangent to the stresses of the principal field, satisfy the differential equation

$$\frac{\partial\Phi}{\partial x}\, dy - \frac{\partial\Phi}{\partial y}\, dx = 0$$

which here becomes

$$2xy\, dy + (3R^2 - x^2 - 3y^2)dx = 0.$$

With an integrating factor of $x^{-4}$ we find as trajectories the curves

$$x^2 + y^2 - R^2 + kx^3 = 0$$

where $k$ is a constant. The curves all pass through the points $x = 0$, $y = \pm R$, where they have horizontal tangents.

For the bending function $\Psi$ of the secondary field, we use relations between the harmonic function

$$Z = \Psi + \mu = \Psi + \frac{1}{6}\nu b(x^3 + 3xy^2),$$

and its conjugate $H$, which at $r = R$ must satisfy

$$\frac{\partial H}{\partial r} = \frac{1}{r}\frac{\partial \mu}{\partial \theta} = \frac{1}{4}\nu bR^2(-\sin\theta + \sin 3\theta).$$

For $H$ we therefore choose a harmonic function of the same type as $M$. After the identifying calculations we have

$$H = \frac{1}{12}\nu b(-3R^2 r\sin\theta + r^3\sin 3\theta)$$

and, therefore,

$$Z = \frac{1}{12}\nu b(3R^2 r\cos\theta - r^3\cos 3\theta),$$

$$\Psi = \frac{1}{4}\nu b(R^2 - r^2)r\cos\theta = \frac{1}{4}\nu b(R^2 - x^2 - y^2)x.$$

The stress trajectories $\Psi = \text{const.}$ show a state of double torsion with the limit points $y = 0$, $x = \pm R/\sqrt{3}$. The problem of warping is thus solved with

$$g = H + 2(1+\nu)\Phi + \frac{1}{3}\nu by^3 = \frac{1}{4}b(3R^2 - x^2 - y^2)y + \frac{1}{2}\nu b\, R^2 y.$$

G. <u>Bending of a circular tube</u>. The boundary conditions for the function $M$ are now

$$\frac{\partial M}{\partial r} = \frac{1}{2}bR^2\sin^3\theta \quad \text{for } r = R, \text{ the exterior circle,}$$

$$\frac{\partial M}{\partial r} = \frac{1}{2}b\rho^2\sin^3\theta \quad \text{for } r = \rho, \text{ the interior circle.}$$

Since the origin is not part of the region D, a pole there is permissible. We set

$$M = \operatorname{Im}(pre^{i\theta}+qr^3e^{3i\theta} + mr^{-1}e^{-i\theta} + nr^{-3}e^{-3i\theta})$$

with p, q, m, and n real, or

$$M = pr \sin\theta + qr^3\sin 3\theta - mr^{-1}\sin\theta - nr^{-3}\sin 3\theta.$$

The boundary conditions determine the parameters:

$$p = \frac{3b}{8}(R^2 + \rho^2), \quad q = -\frac{b}{24}, \quad m = -\frac{3b}{8}\rho^2R^2, \quad n = 0.$$

The principal bending function then becomes

$$\Phi = \frac{b}{24}[9(R^2+\rho^2)r\sin\theta - r^3\sin 3\theta + 9\frac{\rho^2R^2}{r}\sin\theta - 4r^3\sin^3\theta]$$

$$= \frac{b}{8}[3(R^2+\rho^2) - x^2 - y^2 + 3\frac{\rho^2R^2}{x^2+y^2}]y.$$

An analogous procedure for H yields

$$H = \frac{\nu b}{4}[-(R^2+\rho^2)r\sin\theta + \frac{1}{3}r^3\sin 3\theta - R^2\rho^2\frac{\sin\theta}{r}]$$

$$= \frac{\nu b}{4}[-R^2 - \rho^2 + x^2 - \frac{1}{3}y^2 - \frac{R^2\rho^2}{x^2+y^2}]y$$

and, for the conjugate function,

$$Z = \frac{\nu b}{4}[R^2 + \rho^2 + y^2 - \frac{1}{3}x^2 - \frac{R^2\rho^2}{x^2+y^2}]x.$$

The bending function for the secondary field is thus

$$\Psi = \frac{\nu b}{4}(R^2 + \rho^2 - x^2 - y^2 - \frac{R^2\rho^2}{x^2+y^2})x,$$

which vanishes on each of the boundary circles.

The warping is

$$g = \frac{b}{4}(3R^2 + 3\rho^2 - x^2 - y^2 + \frac{3\rho^2R^2}{x^2+y^2})y + \frac{\nu b}{2}(R^2 + \rho^2 + \frac{\rho^2R^2}{x^2+y^2})y.$$

H. Bending of a rectangular bar. The conditions of symmetry are the same as for the circular bar. The bending function for the principal field is simply

$$\Phi = \frac{1}{6} b(-y^3 + 3h^2 y)$$

which satisfies the conditions

$$\nabla^2 \Phi = -by, \quad \frac{\partial \Phi}{\partial x} = 0 \quad \text{for} \quad x = \pm a, \quad \frac{\partial \Phi}{\partial y} = 0 \text{ for } y = \pm h.$$

Indeed $\partial\Phi/\partial x \equiv 0$ and for the principal field we have

$$\tau_{xz} = 0, \quad \tau_{yz} = \frac{1}{2} Eb(h^2 - y^2).$$

For the secondary field, for example, we must find the harmonic function $H$ such that

$$\frac{\partial H}{\partial x} = \frac{\partial \mu}{\partial y} = \nu b\, xy \qquad \text{for} \quad x = \pm a,$$

$$\frac{\partial H}{\partial y} = -\frac{\partial \mu}{\partial x} = -\frac{1}{2} \nu b(x^2 + y^2) \quad \text{for} \quad y = \pm h.$$

Let us try to render one of the two boundary conditions homogeneous and the other well-conditioned from the viewpoint of development in a series of characteristic functions, by adding to $H$ a particular harmonic function. With

$$H = -\frac{1}{6} \nu b(y^3 - 3yx^2 + 6a^2 y) + M$$

the harmonic function $M$ should meet the conditions

$$\frac{\partial M}{\partial x} = 0 \qquad \text{for} \quad x = \pm a,$$

$$\frac{\partial M}{\partial y} = \nu b(a^2 - x^2) \qquad \text{for} \quad y = \pm h.$$

The solution of this problem is similar to that for Prandtl's stress function. $M$ should be even in $x$ and odd in $y$; a solution with variables separated is thus of the type cos $\alpha x$ sinh $\alpha y$. The first boundary condition requires sin $\alpha a = 0$, whence $\alpha = m\,\pi/a$, $m = 0,1,2,\ldots$, and we consider the series

$$\frac{\partial M}{\partial y} = A_o + \sum_1^\infty A_m \cosh \frac{m\pi}{a} y \cos \frac{m\pi}{a} x.$$

The second boundary condition requires

$$A_o + \sum_1^\infty A_m \cosh \frac{m\pi h}{a} \cos \frac{m\pi}{a} x = \nu b(a^2 - x^2);$$

by comparison with the expansion of $\nu b(a^2-x^2)$ we obtain

$$A_o = \tfrac{2}{3}\nu b a^2; \quad A_m \cosh \frac{m\pi h}{a} = 4\nu b a^2 \frac{(-1)^{m+1}}{(m\pi)^2}, \quad m = 1,2,\ldots .$$

By integrating the series with respect to $y$ and again using the first boundary condition we obtain, up to a constant,

$$M = \frac{2}{3}\nu b a^2 y + 4\nu b a^3 \sum_1^\infty \frac{(-1)^{m+1}}{(m\pi)^3} \frac{\sinh \frac{m\pi}{a} y}{\cosh \frac{m\pi h}{a}} \cos \frac{m\pi}{a} x,$$

$$H = -\tfrac{1}{6}\nu b(y^3 - 3yx^2 + 2a^2 y) + 4\nu b a^3 \sum_1^\infty \frac{(-1)^{m+1}}{(m\pi)^3} \frac{\sinh \frac{m\pi}{a} y}{\cosh \frac{m\pi h}{a}} \cos \frac{m\pi}{a} x.$$

One may proceed in the same way for the function $Z$. Write

$$Z = \tfrac{1}{6}\nu b(2a^2 x - x^3 + 3xy^2) + N$$

so that the function $N$ remains harmonic. On the sides of the rectangle $Z$ must equal $\mu$, because $\Psi$ vanishes there. Thus one finds that

$$N = 0 \quad \text{on} \quad x = \pm a, \quad N = \frac{1}{3}\nu b x(x^2 - a^2) \quad \text{on} \quad y = \pm h.$$

Now $N$ is odd in $x$ and even in $y$, whence

$$N = \sum_1^\infty A_m \cosh \frac{m\pi}{a} y \sin \frac{m\pi}{a} x$$

For $y = \pm h$ this should correspond to the Fourier series for $\frac{1}{3}\nu bx(x^2-a^2)$. After calculating the coefficients we find

$$Z = \frac{1}{6}\nu b(2a^2x - x^3 + 3xy^2) + 4\nu ba^3 \sum_1^\infty \frac{(-1)^m}{(m\pi)^3} \frac{\cosh \frac{m\pi}{a} y}{\cosh \frac{m\pi h}{a}} \sin \frac{m\pi}{a} x.$$

It is easy to verify that these functions $H$ and $Z$ satisfy the Cauchy-Riemann equations.

The secondary bending function is

$$\Psi = Z - \mu = \frac{1}{3}\nu bx(a^2-x^2) + 4\nu ba^3 \sum_1^\infty \frac{(-1)^m}{(m\pi)^3} \frac{\cosh \frac{m\pi}{a} y}{\cosh \frac{m\pi h}{a}} \sin \frac{m\pi}{a} x.$$

It always gives the stress trajectorires a configuration of double torsion, symmetric with respect to the plane $x = 0$, on which it also vanishes.

# Chapter 7
# Plane Stress and Plane Strain

The plane problems to be discussed in this chapter occur as exact or approximate solutions of certain three-dimensional problems in the theory of elasticity. For isotropic materials these solutions may be expressed in terms of biharmonic functions of two variables. The use of functions of the corresponding complex variable is clearly indicated, because of the ease with which the solutions can thereby be formed and manipulated.

## 7.1. Lemmas for the Integration of Partial Differential Equations in Complex Form

Let

$$\partial = \frac{\partial}{\partial x} - i \frac{\partial}{\partial y} \tag{7.1}$$

be a complex partial differential operator, and

$$\bar{\partial} = \frac{\partial}{\partial x} + i \frac{\partial}{\partial y} \tag{7.2}$$

the conjugate complex operator. Then for Laplace's operator we have

$$\nabla^2 = \frac{\partial}{\partial x^2} + \frac{\partial}{\partial y^2} = \partial\bar{\partial} = \bar{\partial}\partial. \tag{7.3}$$

Now let

$$f(\zeta) = f(x+iy) = p(x,y) + iq(x,y)$$

be an analytic function of the complex variable $\zeta$. We know that its real part $p(x,y)$ and imaginary part $q(x,y)$ are harmonic and satisfy the Cauchy-Riemann equations

$$\partial p/\partial x = \partial q/\partial y, \quad \partial p/\partial y = -\partial q/\partial x. \tag{7.4}$$

We shall also wish to consider the conjugate analytic function

$$\bar{f}(\zeta) = p(x,y) - iq(x,y).$$

By virtue of the Cauchy-Riemann equations the derivative of the analytic function may take one or another of the following forms:

$$\frac{df}{d\zeta} \equiv f'(\zeta) = \frac{\partial p}{\partial x} + i\,\frac{\partial q}{\partial x} = \frac{\partial q}{\partial y} + i\,\frac{\partial q}{\partial x} = \frac{\partial p}{\partial x} - i\,\frac{\partial p}{\partial y} = \frac{\partial q}{\partial y} - i\,\frac{\partial p}{\partial y}\,.$$

It follows that

$$f'(\zeta) = \partial p = i\partial q,$$

and consequently that

$$\partial f = \partial(p+iq) = 2f'(\zeta), \tag{7.5}$$

$$\partial\bar{f} = \partial(p-iq) = 0. \tag{7.6}$$

More generally, if $f$ and $g$ are two analytic functions,

$$\partial(\bar{g}f) = 2\bar{g}f'. \tag{7.7}$$

Now consider the inverse problem: to find the general integral of

$$\partial h = 2\bar{g}f' \tag{7.8}$$

where the right hand side is given. We find

$$h = \overline{g}f + \overline{k} \tag{7.9}$$

where $k$ is an arbitrary analytic function; indeed, the first term is a particular solution because of (7.7), and by (7.6) the second is the general solution with vanishing right-hand side. If $h = a + ib$ then

$$\partial h = \partial(a+ib) = (\partial a/\partial x + \partial b/\partial y) + i(\partial b/\partial x - \partial a/\partial y),$$

and we see that the problem (7.8) consisted in finding a two-dimensional vector field with specified divergence and curl. If both vanish, then

$$\partial b/\partial x = \partial a/\partial y, \quad \partial b/\partial y = -\partial a/\partial x$$

and the Cauchy-Riemann equations show that $b + ia$ is an analytic function of $\zeta$. Thus $b$ and $a$ will be conjugate harmonic functions and $a+ib = i(b-ia)$ the conjugate of an analytic function. When the data for the divergence and the curl do not vanish, if only they are analytic they may be expanded into Taylor's series which converge (perhaps in a restricted region), and the substitution

$$x = \frac{1}{2}(\zeta + \overline{\zeta}), \quad y = -\frac{i}{2}(\zeta - \overline{\zeta})$$

puts them into the form of a sum of terms of the type appearing in the right-hand side of (7.8).

Now suppose the unknown $h$ is required to be real. In this case the stated problem requires the determination of $h$ from the data of its two partial derivatives $\partial h/\partial x$ and $\partial h/\partial y$. Because $\overline{\partial}\partial h = \overline{\partial}(2\overline{g}f') = \nabla^2 h$ must be real, no choice of the function $\overline{k}$ will cause $h$ to be real unless the data satisfy at the outset the integrability condition*

*Editor's note: In the notation $\overline{f}'$, the derivative is to be taken before the complex conjugate.

$$\text{Im}\,\bar{\partial}(2\bar{g}f') = -\text{Im}\,\partial(2g\bar{f}') = -4\,\text{Im}(g'\bar{f}') = 0.$$

This requires $g' = \alpha f'$ where $\alpha$ is a real constant, so that, up to a constant, the particular integral is $f\bar{f}$ and is real.

## 7.2. The Structure of a Biharmonic Function

Let $h$ be a real biharmonic function of $x$ and $y$, i.e.,

$$\nabla^2\nabla^2 h = \partial\,\bar{\partial}\,\bar{\partial}\,\partial h = 0. \tag{7.10}$$

We may integrate this equation by the preceding method. First

$$\bar{\partial}\,\bar{\partial}\,\partial h = 4\bar{f}'',$$

where the numerical factor and the second derivative of the arbitrary analytic function are only to facilitate later integrations, and do not affect the generality of the result. By taking the conjugate we find

$$\partial\,\partial\,\bar{\partial}h = 4f'',$$

which implies that

$$\partial\,\bar{\partial}h \equiv \nabla^2 h = 2f' + 2\bar{f}'$$

because the result must be real. Then

$$\bar{\partial}h = f + \zeta\bar{f}' + \bar{g}' \quad \text{or} \quad \partial h = \bar{f} + \bar{\zeta}f' + g'$$

and finally

$$h = \frac{1}{2}(\zeta\bar{f} + \bar{\zeta}f) + \frac{1}{2}(g + \bar{g}) = \text{Re}\{\bar{\zeta}f + g\}. \tag{7.11}$$

A general biharmonic function may thus be constructed from two analytic functions. In real terms,

$$h = xp(x,y) + yq(x,y) + r(x,y) \tag{7.11$'$}$$

where $p$ and $q$ are harmonic conjugates and $r$ is harmonic. By taking $\pm\mathrm{Re}[(x+iy)(p+iq)]$ for $r$, one sees that $xp(x,y)$ and $yq(x,y)$ are biharmonic functions. With $f = \zeta k$, we obtain from (7.11) another general representation

$$h = (x^2 + y^2)p(x,y) + r(x,y) \tag{7.11)''}$$

which involves two independent harmonic functions.

## 7.3. Structure of the Solution of the Problems of Plane Strain

The problems of plane strain are strictly two-dimensional from the point of view of the displacements. They correspond to a situation in which

$$w \equiv 0, \quad u = u(x,y), \quad v = v(x,y). \tag{7.12}$$

Because $\varepsilon_z = \partial w/\partial z \equiv 0$, $\gamma_{xz} = \partial u/\partial z + \partial w/\partial x \equiv 0$, and $\gamma_{yz} = \partial u/\partial y + \partial v/\partial x \equiv 0$, the strains reduce to a two-dimensional tensor with components

$$\varepsilon_x = \partial u/\partial x, \quad \varepsilon_y = \partial v/\partial y, \quad \gamma_{xy} = \partial u/\partial y + \partial v/\partial x$$

independent of the $z$-coordinate.

For an isotropic medium one result is

$$\tau_{xz} \equiv 0, \quad \tau_{yz} \equiv 0. \tag{7.13}$$

The non-vanishing stress components $\sigma_x$, $\sigma_y$, $\tau_{xy}$, and $\sigma_z$ are independent of $z$ and are connected with the strains by the constitutive equations

$$E\,\frac{\partial u}{\partial x} = \sigma_x - \nu(\sigma_y+\sigma_z), \quad E\,\frac{\partial v}{\partial y} = \sigma_y - \nu(\sigma_x+\sigma_z), \tag{7.14}$$

$$G\left(\frac{\partial u}{\partial y} + \frac{\partial v}{\partial x}\right) = \tau_{xy}, \tag{7.15}$$

and, because $\varepsilon_z \equiv 0$,

$$\sigma_z = \nu(\sigma_x + \sigma_y). \tag{7.16}$$

With this last relation used to eliminate $\sigma_z$, equations (7.14) take the form

$$\hat{E}\,\frac{\partial u}{\partial x} = \sigma_x - \hat{\nu}\sigma_y, \qquad \hat{E}\,\frac{\partial v}{\partial y} = \sigma_y - \hat{\nu}\sigma_x, \tag{7.17}$$

with the *effective* Young's modulus and Poisson's ratio defined by

$$\hat{E} = \frac{E}{1-\nu^2}, \qquad \hat{\nu} = \frac{\nu}{1-\nu}. \tag{7.18}$$

Under these conditions one finds that the shear modulus is unchanged:

$$\hat{G} = \frac{\hat{E}}{2(1+\hat{\nu})} = \frac{E}{2(1+\nu)} = G.$$

When equations (7.17) are solved for the stresses they yield

$$\sigma_x = 2\hat{G}\left(\frac{\partial u}{\partial x} + \hat{\nu}\hat{\varepsilon}\right), \qquad \sigma_y = 2\hat{G}\left(\frac{\partial v}{\partial y} + \hat{\nu}\hat{\varepsilon}\right) \tag{7.19}$$

where we have set

$$\hat{\varepsilon} = \frac{1}{1-\hat{\nu}}\left(\frac{\partial u}{\partial x} + \frac{\partial v}{\partial y}\right) = \frac{1}{\hat{E}}\,(\sigma_x + \sigma_y). \tag{7.20}$$

This definition needs clarification in the case of an incompressible medium, characterized by $\nu = 0.5$ and consequently $\hat{\nu} = 1$. The first expression for $\hat{\varepsilon}$ becomes indeterminate because $\operatorname{div} u = \partial u/\partial x + \partial v/\partial y = 0$. On the other hand, the second expression remains determinate. One could write $\hat{\varepsilon} = -2p/\hat{E}$ where $p = -(\sigma_x + \sigma_y)/2$ is a pressure which does not depend on a constitutive law (unless the latter is viewed as a passage to the limit) but constitutes a reaction against the geometric condition of incompressibility.

By substituting (7.19) and (7.15) into the equilibrium equations with vanishing body forces, here reduced to

$$\partial\sigma_x/\partial x + \partial\tau_{xy}/\partial y = 0, \quad \partial\tau_{xy}/\partial x + \partial\sigma_y/\partial y = 0, \tag{7.21}$$

one finds Navier's equations for the problem. They can be presented in the form of Cauchy-Riemann equations

$$\partial\hat{\varepsilon}/\partial x = \partial\omega/\partial y, \quad \partial\hat{\varepsilon}/\partial y = -\partial\omega/\partial x, \tag{7.22}$$

where

$$\omega = \frac{1}{2}\left(\frac{\partial v}{\partial x} - \frac{\partial u}{\partial y}\right) \tag{7.23}$$

denotes the material rotation of the fibers parallel to Oz. We may then write

$$\hat{\varepsilon} + i\omega = 2F'(\zeta)/\hat{G} \tag{7.24}$$

where $F(\zeta)$ is an analytic function of a complex variable $\zeta = x + iy$, and we conclude that

$$\frac{\partial u}{\partial x} + \frac{\partial v}{\partial y} = (1-\hat{\nu})\hat{\varepsilon} = \frac{1-\hat{\nu}}{\hat{G}}(F' + \overline{F}'),$$

$$\frac{\partial v}{\partial x} - \frac{\partial u}{\partial y} = 2\omega = \frac{2}{i\hat{G}}(F' - \overline{F}').$$

These two equations may be rewritten in the complex form

$$\partial(u+iv) = (1-\hat{\nu})\hat{\varepsilon} + 2i\omega = \{(3-\hat{\nu})F' - (1+\hat{\nu})\overline{F}'\}/\hat{G}.$$

Application of the result on integration then yields

$$u + iv = \{(3-\hat{\nu})F - (1+\hat{\nu})\zeta\overline{F}' + \overline{H}'\}/2\hat{G}, \tag{7.25}$$

and the displacement field is expressed in terms of two analytic functions $F(\zeta)$ and $H(\zeta)$.

The stress field follows without difficulty. From (7.24) and (7.20) we deduce

$$\sigma_x + \sigma_y + i\hat{E}\omega = 4(1+\hat{\nu})F'. \tag{7.26}$$

Then, by combining (7.19) and (7.15), we find

$$\sigma_x - \sigma_y - 2i\tau_{xy} = 2\hat{G}\partial(u-iv)$$

which becomes, after using (7.25),

$$\sigma_x - \sigma_y - 2i\tau_{xy} = -2(1+\hat{\nu})\bar{\zeta}F'' + 2H''. \tag{7.27}$$

Finally, $\sigma_z$ follows directly from (7.16).

## 7.4. Structure of the Solution of the Problem of Plane Stress

We consider here a plate with middle surface at $z = 0$ and traction-free bounding faces at $z = \pm h$. Since $\sigma_z$, $\tau_{zx}$, and $\tau_{zy}$ then vanish on the faces, one may seek a solution corresponding to a state of *plane stress*, that is, one for which

$$\sigma_z \equiv 0, \quad \tau_{xz} \equiv 0, \quad \tau_{yz} \equiv 0. \tag{7.28}$$

The equilibrium equations with vanishing body forces are identical with (7.21) governing a state of plane strain, and three of the constitutive relations are similar, viz., (7.15) and

$$E\,\frac{\partial u}{\partial x} = \sigma_x - \nu\sigma_y, \qquad E\,\frac{\partial v}{\partial y} = \sigma_y - \nu\sigma_x, \tag{7.17'}$$

which require only suitable adjustments of Young's modulus and Poisson's ratio. Aside from this last difference, there is a more fundamental one: if one wishes an exact solution in the context of three-dimensional elasticity, it is necessary to admit displacements and stresses which vary with $z$. This requires us to consider also the relations

$$\tau_{xz}/G = \partial u/\partial z + \partial w/\partial x = 0, \quad \tau_{yz}/G = \partial v/\partial z + \partial w/\partial y = 0, \tag{7.29}$$

(which were seen to be satisfied identically in the preceeding section), along with

$$E\,\frac{\partial w}{\partial z} = -\nu(\sigma_x + \sigma_y). \tag{7.30}$$

Subtracting this last equation from each of the equations (7.17)' yields

$$\sigma_x = 2G(\frac{\partial u}{\partial x} - \frac{\partial w}{\partial z}), \quad \sigma_y = 2G(\frac{\partial v}{\partial y} - \frac{\partial w}{\partial z}), \tag{7.31}$$

which compare with (7.19). With this result and (7.15) substituted into the equilibrium equations (7.21), Navier's equations again take a Cauchy-Riemann form

$$\partial\varepsilon/\partial x = \partial\omega/\partial y, \quad \partial\varepsilon/\partial y = -\partial\omega/\partial x,$$

where now $\varepsilon$ is defined by

$$\varepsilon = \partial u/\partial x + \partial v/\partial y - \partial w/\partial z. \tag{7.32}$$

The analytic function $\varepsilon + i\omega$ of the variable $\zeta = x + iy$ now contains the $z$-coordinate as parameter. Eliminating $w$ from equations (7.29) yields

$$\partial\omega/\partial z = 0. \tag{7.33}$$

If the imaginary part $\omega$ is thus independent of $z$, the real part must be the sum of a function of $z$ alone and a harmonic function independent of $z$. Consequently, one writes

$$\varepsilon + i\omega = 2\{a(z) + F'(\zeta)\}/G \tag{7.34}$$

where $a(z)$ is a real function of $z$.

By adding equations (7.17)' there follows

$$E(\partial u/\partial x + \partial v/\partial y) = (1-\nu)(\sigma_x+\sigma_y).$$

Subtraction of (7.30) from this yields

$$E\varepsilon = \sigma_x + \sigma_y. \tag{7.35}$$

From this we conclude

$$\frac{\partial u}{\partial x} + \frac{\partial v}{\partial y} = \frac{1-\nu}{E}(\sigma_x + \sigma_y) = (1-\nu)\varepsilon,$$

and therefore

$$\partial(u+iv) = (1-\nu)\varepsilon + 2i\omega = 2\{(1-\nu)a + \frac{1-\nu}{2}(F'+\overline{F}')+F'-\overline{F}'\}/G,$$

of which the general integral will be

$$u+iv = \{2(1-\nu)a\zeta + (3-\nu)F - (1+\nu)\zeta\overline{F}' + \overline{K}(\zeta;z)\}/2G.$$

The nature of the $z$ dependence is found by studying the integrability conditions of the displacement $w$. Equation (7.33) is one, and it is already satisfied. The other two result from the elimination of $w$ among equations (7.29) and

$$\partial w/\partial z = -\nu\varepsilon, \tag{7.36}$$

the latter being a consequence of (7.30) and (7.35). The result may be put into the complex form

$$\frac{\partial^2}{\partial z^2}(u - iv) = \nu\partial\varepsilon.$$

After inserting the solutions found for $u$, $v$, and $\varepsilon$, the condition becomes

$$2(1-\nu)\overline{\zeta}\,\frac{d^2a}{dz^2} + \frac{\partial^2 K}{\partial z^2} = 4\nu F'',$$

which separates into

$$d^2a/dz^2 = 0, \qquad \partial^2K/\partial z^2 = 4\nu F''.$$

We choose integrals which satisfy the symmetry conditions

$$u(x,y,-z) = u(x,y,+z), \quad v(x,y,-z) = v(x,y,+z).$$

These characterize an extension of the plate, and deliberately avoid a solution representing bending of the plate by bending moments.

For $K$, for example, we have

$$K = 2\nu(z^2 - h^2/3)F'' + H'$$

where $H$ is an analytic function of $\zeta$ only. The real function $a(z)$, which should be even, can only be a constant, and there is no loss in assuming it absorbed into $H'$. Then

$$u+iv = \{(3-\nu)F - (1+\nu)\zeta\overline{F}' + 2\nu(z^2 - \tfrac{1}{3}h^2)\overline{F}'' + \overline{H}'\}/2G. \quad (7.37)$$

Now we find the transverse displacement by putting equations (7.29) into their complex form

$$\partial w = -\frac{\partial}{\partial z}(u-iv) = -\frac{2\nu}{G} zF''.$$

The integral satisfying (7.36) is

$$w = -\frac{\nu z}{G}(F' + \overline{F}') = -\frac{2\nu z}{G}\,\mathrm{Re}\,F'. \quad (7.38)$$

For the stresses we have

$$\sigma_x + \sigma_y + iE\omega = E(\varepsilon + i\omega) = 4(1+\nu)F', \quad (7.39)$$

$$\sigma_x - \sigma_y - 2i\tau_{xy} = 2G\partial(u-iv) = -2(1+\nu)\overline{\zeta}F'' + 4\nu(z^2 - \tfrac{1}{3}h^2)F''' + 2H''. \quad (7.40)$$

## 7.5. Generalized Plane Stress

In *Love's* terminology, the problem treated in the preceding section is that of plane stress. There we sought the form of an exact solution of three-dimensional elasticity, with the properties (7.28) reducing the stress tensor to a two-dimensional structure which still depends partially on the third coordinate. The *generalized* state of plane stress is not usually an exact solution of three-dimensional elasticity. It corresponds to the plane state obtained by substituting for the quantities $(u,v)$ and $(\sigma_x, \tau_{xy}, \sigma_y)$ their averages over the thickness of the plate. Since

$$\int_{-h}^{h} (z^2 - \frac{1}{3} h^2) dz = 0,$$

it is enough to suppress the terms in $(z^2 - \frac{1}{3} h^2)$ in the results of Section 7.4. The solution for generalized plane stress thus takes the form

$$u + iv = \{(3-\nu)F - (1+\nu)\zeta\overline{F}' + H'\}/2G, \tag{7.41}$$

$$\sigma_x + \sigma_y + iE\omega = 4(1+\nu)F', \tag{7.42}$$

$$\sigma_x - \sigma_y - 2i\tau_{xy} = -2(1+\nu)\overline{\zeta}F'' + 2H''. \tag{7.43}$$

The analogy with the case of plane strain is now clear. It is enough to compare equations (7.41)-(7.43) with (7.25)-(7.27) to see that the results can be deduced from each other by the correspondences $E \leftrightarrow \hat{E}$ and $\nu \leftrightarrow \hat{\nu}$. It should be remembered, however, that while $\sigma_z = 0$ in the state of plane stress, with $\varepsilon_z \neq 0$, this situation is reversed in the state of plane strain. The results (7.37) to (7.40) generalize those obtained by *Kolosov* and *Mushkelishvili*; the terms they contain in $(z^2 - h^2/3)$ are proportional to Poisson's ratio and can have important effects in regions of high gradients of $F'$ for the displacements or $F''$ for the stresses.

The determination of the functions $F$ and $H$ is obviously connected with the boundary conditions of the problem. One usually distinguishes the case in which the surface tractions are specified on the boundaries, called the first fundamental problem, from the case in which the displacements are specified, called the second fundamental problem.*

When the region extends to infinity it will be necessary to specify the behavior of the stresses there, as well as the material rotation and the resultant of the applied

*Editor's note: This is the reverse of most writers' usage.

exterior forces.

## 7.6. Airy's Stress Function

Let O be the origin and P(x,y) be an arbitrary point of the middle surface of a plate. A path $c_1$ drawn from O to P in the middle surface defines a cylindrical strip with generators parallel to the axis Oz, which we shall regard as an exterior face with outer normal situated, for example, to the right with respect to the direction of traversal.

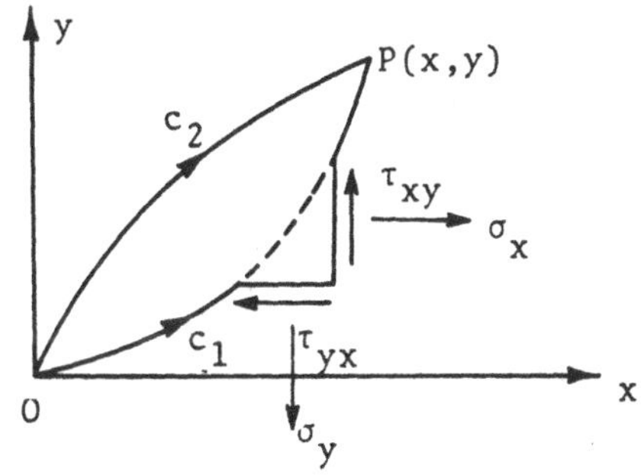


Figure 7.1

The resultant of the surface tractions on this face of the strip includes the components of a force per unit thickness

$$X_{c_1} = \int_{c_1} \sigma_x dy - \tau_{yx} dx, \quad Y_{c_1} = \int_{c_1} \tau_{xy} dy - \sigma_y dx, \tag{7.44}$$

and a moment with respect to the Oz axis

$$M_{c_1} = \int_{c_1} (x\tau_{xy} - y\sigma_x) dy - (x\sigma_y - y\tau_{yx}) dx. \tag{7.45}$$

If the slab is simply connected, every other path $c_2$ drawn from O to P in the medium is reconcilable with the first (cf. Section 5.2) or, in other words, the pair of paths forms a reducible closed circuit. In the absence of body forces and surface tractions on the faces $z = \pm h$, the equilibrium of the portion of the slab inside the closed circuit is expressed by

$$X_{c_1} - X_{c_2} = 0, \quad Y_{c_1} - Y_{c_2} = 0, \quad M_{c_1} - M_{c_2} = 0,$$

because the exterior normal undergoes a change in orientation with respect to the direction of passage from O to P when one goes from one path to the other. These relations say that the quantities (7.44) and (7.45) are independent of the path traversed from O to P(x,y). The integrands are thus exact differentials of univalent functions. Thus for (7.44) we have

$$\sigma_x dy - \tau_{yx} dx = dX, \quad \tau_{xy} dy - \sigma_y dx = dY, \tag{7.46}$$

which imply the relations

$$\sigma_x = \frac{\partial X}{\partial y}, \quad \tau_{yx} = -\frac{\partial X}{\partial x}, \quad \tau_{xy} = \frac{\partial Y}{\partial y}, \quad \sigma_y = -\frac{\partial Y}{\partial x}. \tag{7.47}$$

Similarly, from (7.45)

$$(x\tau_{xy} - y\sigma_x)dy - (x\sigma_y - y\tau_{yx})dx = dM$$

and, if we substitute formulas (7.47),

$$x(\frac{\partial Y}{\partial y} dy + \frac{\partial Y}{\partial x} dx) - y(\frac{\partial X}{\partial y} dy + \frac{\partial X}{\partial x} dx) = xdY - ydX = dM.$$

By setting

$$\phi = M - xY + yX \tag{7.48}$$

this may be written

$$d\phi = d(M - xY + yX) = X\,dy - Y\,dx \tag{7.49}$$

and is equivalent to the relations

$$X = \partial\phi/\partial y, \quad Y = -\partial\phi/\partial x. \tag{7.50}$$

These transform formulas (7.47) into

$$\sigma_x = \frac{\partial^2\phi}{\partial y^2}, \quad \tau_{xy} = \tau_{yx} = -\frac{\partial^2\phi}{\partial x\partial y}, \quad \sigma_y = \frac{\partial^2\phi}{\partial x^2}. \tag{7.51}$$

The function $\phi(x,y)$ is known as *Airy's stress function*. It is constructed so as to satisfy the equilibrium equations with vanishing body forces

$$\partial\sigma_x/\partial x + \partial\tau_{yx}/\partial y = 0, \quad \partial\tau_{xy}/\partial x + \partial\sigma_y/\partial y = 0, \tag{7.52}$$

as may be verified at once.

This derivation immediately furnishes a physical interpretation of $\phi$ and its partial derivatives. By integrating the formulas (7.46) from O to P we first find the resultant force components of the surface tractions to be

$$\begin{aligned} X_P &= X(x,y) - X(0,0) = \frac{\partial\phi}{\partial y} - \left(\frac{\partial\phi}{\partial y}\right)_0 \\ X_P &= Y(x,y) - Y(0,0) = -\frac{\partial\phi}{\partial x} + \left(\frac{\partial\phi}{\partial x}\right)_0. \end{aligned} \tag{7.53}$$

Then the integration of (7.49) gives the moment with respect to the origin $M_0$ as

$$\phi(x,y) - \phi(0,0) = M_0 - xY(x,y) + yX(x,y)$$

or

$$M_0 = \phi(x,y) - \phi(0,0) - x\frac{\partial\phi}{\partial x} - y\frac{\partial\phi}{\partial y}, \tag{7.54}$$

and the moment with respect to the point P as

$$M_P = M_0 - xY_P + yX_P = \phi(x,y) - \phi(0,0) - x\left(\frac{\partial\phi}{\partial x}\right)_0 - y\left(\frac{\partial\phi}{\partial y}\right)_0. \tag{7.54'}$$

Equations (7.51) show that the stress field is unchanged when one adds to $\phi$ an arbitrary linear form $\alpha + \beta x + \gamma y$. The coefficients may be chosen so as to give $\phi$ and its partial derivatives arbitrarily chosen values at an arbitrarily chosen point. It is thus permissible to take $\phi$ and its partial derivatives to vanish at the origin, in which case (7.53) and (7.54) simplify to

$$X_P = \frac{\partial\phi}{\partial y}, \quad Y_P = -\frac{\partial\phi}{\partial x}, \quad M_P = \phi. \tag{7.55}$$

We will sometimes call $\alpha$, $\beta$, and $\gamma$, or their equivalents in polar coordinates, the *unproductive parameters* of Airy's function.

When the plate has cavities, so that it represents a multiply connected region, Airy's function and its first partial derivatives may be multivalued functions.

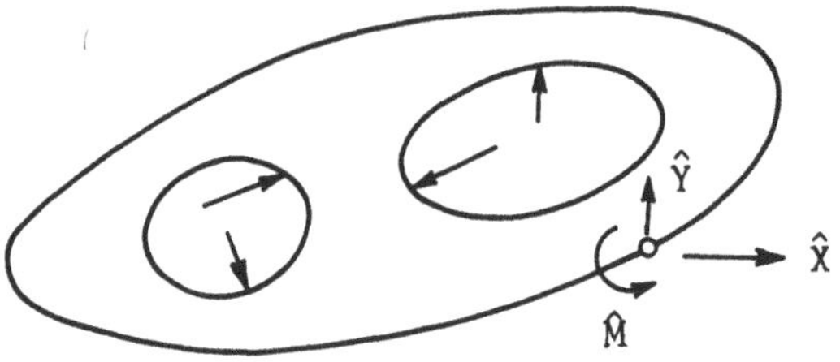


Figure 7.2

The equilibrium of a portion of the plate bounded by an irreducible closed circuit requires

$$\oint dX + \hat{X} = 0 \qquad \oint dY + \hat{Y} = 0 \qquad \oint dM + \hat{M}_o = 0$$

where the integrals represent the resultant of the surface tractions acting along the closed circuit, and $(\hat{X},\hat{Y},\hat{M})$ that of the tractions applied along the edges of the surrounded cavities. Then (7.50) yields the change in the partial derivatives of Airy's stress function upon traversing the closed circuit once:

$$\Delta\frac{\partial\phi}{\partial y} = \oint d\,\frac{\partial\phi}{\partial y} = -\hat{X}, \qquad \Delta\frac{\partial\phi}{\partial x} = \oint d\,\frac{\partial\phi}{\partial x} = \hat{Y}. \tag{7.56}$$

Formula (7.49) allows the same calculation for Airy's function itself, yielding

$$\Delta\phi = \oint d\phi = \oint dM + \oint d(yX - xY) = -\hat{M}_o + \Delta(yX - xY),$$

which leads to

$$\Delta\phi = -\hat{M}_0 - y\hat{X} + x\hat{Y} = -\hat{M}(x,y) \tag{7.57}$$

where $(x,y)$ is an arbitrarily chosen point of the path and $\hat{M}(x,y)$ the moment of the cavity forces with respect to that point. Thus $\phi$ and its first derivatives remain univalent if the cavities are not loaded or, more generally, if for each cavity the applied tractions are statically equivalent to zero.

## 7.7. Complex Representation of Airy's Function

Airy's function has the dimensions of a moment of force per unit thickness. When the medium is isotropic it can be expressed equally well by using the functions $F(\zeta)$ and $H(\zeta)$. By comparing (7.51) with (7.42) and (7.43) we obtain

$$\sigma_x + \sigma_y = \nabla^2\phi = \partial\bar{\partial}\phi = 2(1+\nu)(F' + \bar{F}') \tag{7.58}$$

$$\begin{aligned} \sigma_x - \sigma_y - 2i\tau_{xy} &= \frac{\partial^2\phi}{\partial y^2} - \frac{\partial^2\phi}{\partial x^2} + 2i\,\frac{\partial^2\phi}{\partial x\partial y} \\ &= -\partial\partial\phi = -2(1+\nu)\bar{\zeta}F'' + 2H''. \end{aligned} \tag{7.59}$$

Integration of the last equation yields

$$\partial\phi = (1+\nu)\bar{\zeta}F' - H' + \bar{K}$$

where $K(\zeta)$ is a new unknown analytic function. Pass to the complex conjugate relation

$$\bar{\partial}\phi = (1+\nu)\zeta\bar{F}' - \bar{H}' + K$$

and apply to it the operator $\partial$ to obtain

$$\partial\bar{\partial}\phi = 2(1+\nu)\bar{F}' + 2K'.$$

By comparing with (7.58) one sees that $K' = (1+\nu)F'$ and consequently

$$\partial\phi = (1+\nu)\bar{\zeta}F' + (1+\nu)\bar{F} - H'. \tag{7.60}$$

A final integration yields

$$\phi = \frac{1+\nu}{2}(\bar{\zeta}F + \zeta\bar{F}) - \frac{1}{2}(H + \bar{H}). \tag{7.61}$$

This expression depends in fact only on the two analytic functions $(1+\nu)F(\zeta)$ and $H(\zeta)$. For the case of plane stress, the same integration process gives a supplementary term of three-dimensional effect:

$$\phi = \frac{1+\nu}{2}(\bar{\zeta}F + \zeta\bar{F}) - \nu(z^2 - \frac{1}{3}h^2)(F' + \bar{F}') - \frac{1}{2}(H + \bar{H}). \tag{7.62}$$

For plane strain, the integration leads to (7.61) with $\hat{\nu}$ instead of $\nu$.

By referring to (7.11) we see that Airy's stress function is always biharmonic. This is also a direct consequence of the Beltrami-Michell equations. Indeed, they require that the first invariant $\sigma_x + \sigma_y + \sigma_z$ be a harmonic function of $(x,y,z)$. In the plane strain problem, from (7.16) and the fact that the stresses are independent of $z$ one obtains

$$\left(\frac{\partial^2}{\partial x^2} + \frac{\partial^2}{\partial y^2}\right)(\sigma_x + \sigma_y) = \nabla^2(\sigma_x + \sigma_y) = \nabla^2\nabla^2\phi = 0.$$

In the case of generalized plane stress the same conclusion follows from the facts that $\sigma_z = 0$ and that $\partial^2(\sigma_x+\sigma_y)/\partial z^2 = 0$.

## 7.8. Polar Coordinates

The complex representation allows an easy passage from Cartesian coordinates $(x,y)$ to polar coordinates $(r,\theta)$. Before making this change, a preliminary formulation seems useful, drawing upon basic principles to obtain the fundamental kinematic, static, and constitutive equations.

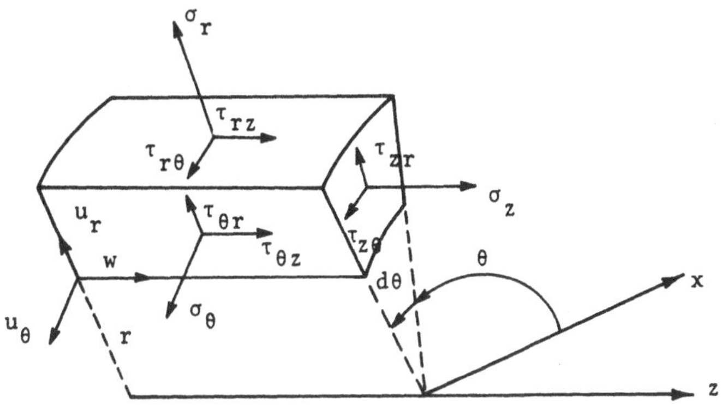


Figure 7.3

Let us calculate the virtual work performed by the stresses on an element with coordinates between $r$ and $r + dr$, $\theta$ and $\theta + d\theta$, and $z$ and $z + dz$. In accordance with the principle of geometric linearization, we may work in the reference configuration, and we find for the virtual work the expression

$$\frac{\partial}{\partial r}\{r\,d\theta\,dz(\sigma_r\,\delta u_r + \tau_{r\theta}\,\delta u_\theta + \tau_{rz}\,\delta w)\}dr$$

$$+ \frac{\partial}{\partial\theta}\{dr\,dz(\tau_{\theta r}\,\delta u_r + \sigma_\theta\,\delta u_\theta + \tau_{\theta z}\,\delta w)\}d\theta$$

$$+ \frac{\partial}{\partial z}\{r\,d\theta dr(\tau_{zr}\,\delta u_r + \tau_{z\theta}\,\delta u_\theta + \sigma_z\,\delta w\}dz.$$

This three-dimensional calculation, using in fact a system of cylindrical coordinates, reduces to a calculation in polar coordinates in the case of plane strain of an isotropic medium where, according to (7.12) and (7.13),

$$\delta w \equiv 0, \quad \tau_{rz} = \cos\theta\,\tau_{xz} + \sin\theta\,\tau_{yz} \equiv 0,$$

$$\tau_{\theta z} = -\sin\theta\,\tau_{xz} + \cos\theta\,\tau_{yz} \equiv 0.$$

The same holds true in the state of plane stress, when (7.28) applies. In each case the only surviving terms are

$$[\frac{\partial}{\partial r}(r\sigma_r\ \delta u_r + r\tau_{r\theta}\ \delta u_\theta) + \frac{\partial}{\partial\theta}(\tau_{\theta r}\ \delta u_r + \sigma_\theta\ \delta u_\theta)]dr\ d\theta\ dz$$

$$= \mathrm{Re}[\frac{\partial}{\partial r}\ \{r(\sigma_r - i\tau_{r\theta})\delta(u_r + iu_\theta)\}$$

$$+ \frac{\partial}{\partial\theta}\ \{(\tau_{\theta r} - i\sigma_\theta)\delta(u_r + iu_\theta)\}]dr\ d\theta\ dz.$$

Dividing this expression by the element of volume $r\ d\theta\ dr\ dz$ and adding the virtual work of the body forces yields the virtual work per unit initial volume:

$$\delta W = \mathrm{Re}[\frac{1}{r}\ \frac{\partial}{\partial r}\{r(\sigma_r - i\tau_{r\theta})\delta(u_r + iu_\theta)\} + \frac{1}{r}\ \frac{\partial}{\partial\theta}\{(\tau_{\theta r} - i\sigma_\theta)\delta(u_r + iu_\theta)\}$$

$$+ \rho_o(g_r - ig_\theta)\delta(u_r + iu_\theta)]. \qquad (7.63)$$

As in the general case, the equilibrium equations arise by expressing the vanishing of the virtual work in a change of configuration compatible with a rigid element. A translation of the element, for example, is expressed in the form

$$\delta(u_r + iu_\theta) = e^{-i\theta}\ \delta(u + iv)$$

where $\delta u$ and $\delta v$ are two arbitrary constants. In these conditions annulling the virtual work amounts to annulling the real part of

$$e^{-i\theta}\delta(u+iv)[\frac{\sigma_r - i\tau_{r\theta}}{r} + \frac{\partial}{\partial r}(\sigma_r - i\tau_{r\theta}) - \frac{i}{r}(\tau_{\theta r} - i\sigma_\theta)$$

$$+ \frac{1}{r}\ \frac{\partial}{\partial\theta}(\tau_{\theta r} - i\sigma_\theta) + \rho_o(g_r - ig_\theta)]. \qquad (7.64)$$

However, since the factor preceding the bracket is an arbitrary complex quantity, the real and imaginary parts inside the bracket should vanish separately. This furnishes the translational equilibrium equations

$$\frac{\partial \sigma_r}{\partial r} + \frac{\sigma_r - \sigma_\theta}{r} + \frac{1}{r}\frac{\partial \tau_{\theta r}}{\partial \theta} + \rho_0 g_r = 0,$$
$$\frac{\partial \tau_{r\theta}}{\partial r} + \frac{\tau_{r\theta} + \tau_{\theta r}}{r} + \frac{1}{r}\frac{\partial \sigma_\theta}{\partial \theta} + \rho_0 g_\theta = 0. \tag{7.65}$$

Several terms in (7.63) now disappear because of the vanishing of the bracket of (7.64); we are left with

$$\delta W = \mathrm{Re}[\tfrac{i}{r}(\tau_{\theta r} - i\sigma_\theta)\delta(u_r + iu_\theta) + (\sigma_r - i\tau_{r\theta})\tfrac{\partial}{\partial r}(\delta u_r + i\delta u_\theta)$$
$$+ (\tau_{\theta r} - i\sigma_\theta)\,\frac{1}{r}\,\frac{\partial}{\partial \theta}(\delta u_r + i\delta u_\theta)]. \tag{7.66}$$

Consider now an infinitesimal virtual rigid rotation of the element characterized by

$$\delta u_r = 0, \qquad \delta u_\theta = r\delta\alpha.$$

Since the virtual work is again zero, the coefficient of the arbitrary constant $\delta\alpha$ must vanish, leading to

$$-\tau_{\theta r} + \tau_{r\theta} = 0 \tag{7.67}$$

which is the equation of rotational equilibrium.

With (7.66) now in the canonical form

$$\delta W = \sigma_r\ \delta\varepsilon_r + \tau_{r\theta}\ \delta\gamma_{r\theta} + \sigma_\theta\ \delta\varepsilon_\theta, \tag{7.68}$$

by identifying the coefficients of $\sigma_r$, $\tau_{r\theta}$, and $\sigma_\theta$ we obtain the kinematic relations

$$\varepsilon_r = \frac{\partial u_r}{\partial r}, \qquad \varepsilon_\theta = \frac{u_r}{r} + \frac{1}{r}\frac{\partial u_\theta}{\partial \theta}$$
$$\gamma_{r\theta} = -\frac{u_\theta}{r} + \frac{\partial u_\theta}{\partial r} + \frac{1}{r}\frac{\partial u_r}{\partial \theta}. \tag{7.69}$$

If the strain energy density $W$ is known as a function of the strains $(\varepsilon_r, \gamma_{r\theta}, \varepsilon_\theta)$, then expression (7.68) is equivalent to the general constitutive equations

$$\sigma_r = \partial W/\partial\varepsilon_r, \quad \tau_{r\theta} = \partial W/\partial\gamma_{r\theta}, \quad \sigma_\theta = \partial W/\partial\varepsilon_\theta. \qquad (7.70)$$

For a state of plane strain, we find

$$W = \frac{1}{2} K(\varepsilon_r+\varepsilon_\theta)^2 + G\{\tfrac{2}{3}(\varepsilon_r^2 + \varepsilon_\theta^2 - \varepsilon_r\varepsilon_\theta) + \frac{1}{2} \gamma_{r\theta}^2\}.$$

This follows from the general form (5.67) by calculating the invariants $I_1$ and $-\Gamma_2$, using the fact that the strain tensor takes the form

$$\begin{pmatrix} \varepsilon_r & \gamma_{r\theta}/2 & 0 \\ \gamma_{r\theta}/2 & \varepsilon_\theta & 0 \\ 0 & 0 & 0 \end{pmatrix}$$

in the locally orthogonal system of axes $\mathbf{e}_r$, $\mathbf{e}_\theta$, $\mathbf{e}_z$. With Lamé's parameters one obtains the simple formulas

$$\sigma_r = \lambda(\varepsilon_r + \varepsilon_\theta) + 2\mu\varepsilon_r, \qquad \sigma_\theta = \lambda(\varepsilon_r + \varepsilon_\theta) + 2\mu\varepsilon_\theta,$$

$$\tau_{r\theta} = \mu\gamma_{r\theta}.$$

We see that the relation

$$\sigma_z = \nu(\sigma_r + \sigma_\theta)$$

cannot be obtained by considerations of virtual work, this stress not performing any virtual work as long as $w \equiv 0$.

For problems of plane stress, we find

$$W = \frac{1}{2} \hat{E}(\varepsilon_r^2 + \varepsilon_\theta^2 + 2\nu\varepsilon_r\varepsilon_\theta) + \frac{1}{2} G\gamma_{r\theta}^2.$$

This is obtained most easily by specializing Hooke's law according to the hypotheses (7.28):

$$E\varepsilon_r = \sigma_r - \nu\sigma_\theta, \quad E\varepsilon_\theta = \sigma_\theta - \nu\sigma_r, \quad G\gamma_{r\theta} = \tau_{r\theta}.$$

When these equations are solved for the stresses and substituted into Clapeyron's formula $W = (\sigma_r\varepsilon_r+\sigma_\theta\varepsilon_\theta+\tau_{r\theta}\gamma_{r\theta})/2$ the stated expression results. The additional relation

$$\varepsilon_z = -\hat{\nu}(\varepsilon_r + \varepsilon_\theta)$$

cannot be obtained by considerations of virtual work, but is a direct consequence of Hooke's law

$$E\varepsilon_z = -\nu(\sigma_r + \sigma_\theta).$$

As in the case of Cartesian coordinates, the equilibrium equations with vanishing body forces (7.65) can be satisfied by using an Airy's stress function. Let us calculate the virtual work of tractions on the surface of unit thickness defined by a closed contour:

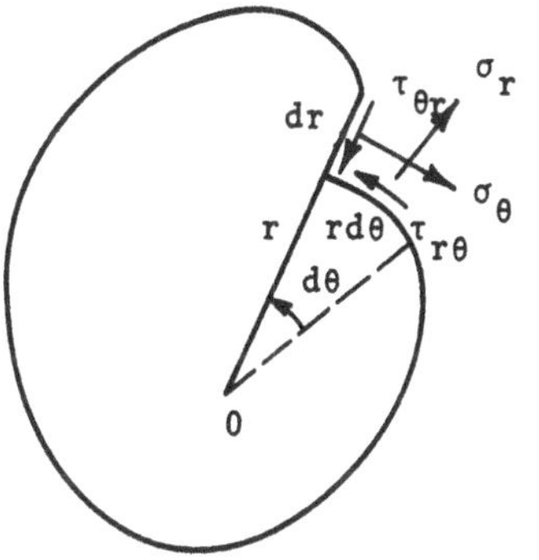


Figure 7.4

$$\oint (\sigma_r \delta u_r + \tau_{r\theta}\delta u_\theta) r d\theta - (\tau_{\theta r}\delta u_r + \sigma_\theta \delta u_\theta) dr$$

$$= \text{Re} \oint r d\theta (\sigma_r - i\tau_{r\theta})\delta(u_r + iu_\theta) - dr(\tau_{\theta r} - i\sigma_\theta)\delta(u_r + iu_\theta).$$

In the absence of body forces, with the contour surrounding no loaded cavities this virtual work must vanish when the displacements represent a rigid body translation of the element. Thus for

$$\delta u_r + i\delta u_\theta = e^{-i\theta}(\delta u + i\delta v)$$

we have

$$\oint e^{-i\theta}\{(\sigma_r - i\tau_{r\theta})r d\theta - (\tau_{\theta r} - i\sigma_\theta)dr\} = 0.$$

The integrand must be the exact differential of a univalent function. Let us set

$$e^{-i\theta}\{(\sigma_r - i\tau_{r\theta})r d\theta - (\tau_{\theta r} - i\sigma_\theta)dr\} = d\{e^{-i\theta}(A+iB)\}.$$

Then

$$r\sigma_r = \frac{\partial A}{\partial\theta} + B, \quad r\tau_{r\theta} = -\frac{\partial B}{\partial\theta} + A, \quad \tau_{\theta r} = -\frac{\partial A}{\partial r}, \ \sigma_\theta = \frac{\partial B}{\partial r}.$$

The equilibrium equations (7.67) then require

$$\frac{A}{r} + \frac{\partial A}{\partial r} - \frac{1}{r}\frac{\partial B}{\partial\theta} = 0 \quad \text{or} \quad \frac{\partial}{\partial r}(rA) = \frac{\partial B}{\partial\theta}$$

and this equation is satisfied by choosing

$$rA = \partial\phi/\partial\theta, \quad B = \partial\phi/\partial r.$$

The final result is

$$\begin{aligned} \sigma_r &= \frac{1}{r^2}\frac{\partial^2\phi}{\partial\theta^2} + \frac{1}{r}\frac{\partial\phi}{\partial r}, \qquad \sigma_\theta = \frac{\partial^2\phi}{\partial r^2}, \\ \tau_{r\theta} &= \tau_{\theta r} = -\frac{\partial}{\partial r}\left(\frac{1}{r}\frac{\partial\phi}{\partial\theta}\right). \end{aligned} \tag{7.71}$$

In the locally orthogonal frame $\mathbf{e}_r, \mathbf{e}_\theta, \mathbf{e}_z$ the stress tensor takes the form

$$\begin{pmatrix} \sigma_r & \tau_{r\theta} & 0 \\ \tau_{\theta r} & \sigma_\theta & 0 \\ 0 & 0 & \sigma_z \end{pmatrix}$$

for all cases discussed, with $\sigma_z$ vanishing in generalized plane stress or plane stress, and in the latter case the other stresses varying with $z$.

In the passage from Cartesian to polar (cylindrical) coordinates the tensor transformation

$$\begin{aligned}\sigma_r &= \cos^2\theta\ \sigma_x + \sin^2\theta\ \sigma_y + 2\sin\theta\cos\theta\ \tau_{xy},\\ \sigma_\theta &= \sin^2\theta\ \sigma_x + \cos^2\theta\ \sigma_y - 2\sin\theta\cos\theta\ \tau_{xy},\\ \tau_{r\theta} &= (\cos^2\theta - \sin^2\theta)\tau_{xy} + (\sigma_y-\sigma_x)\sin\theta\cos\theta,\end{aligned} \tag{7.72}$$

is equivalent with the relations

$$\sigma_r + \sigma_\theta = \sigma_x + \sigma_y \tag{7.73}$$

$$\sigma_r - \sigma_\theta - 2i\tau_{r\theta} = e^{2i\theta}(\sigma_x-\sigma_y-2i\tau_{xy}). \tag{7.74}$$

Also, for the displacements,

$$u_r + iu_\theta = e^{-i\theta}(u + iv). \tag{7.75}$$

These relations allow the immediate transformation of the solutions of plane problems from their Cartesian to their polar structure. For the problem of plane strain,

$$\begin{aligned}u_r + iu_\theta &= \frac{e^{-i\theta}}{2\hat{G}}\{(3-\hat{\nu})F - (1+\hat{\nu})\zeta\overline{F}' + \overline{H}'\},\\ \sigma_r + \sigma_\theta + i\hat{E}\omega &= 4(1+\hat{\nu})F',\\ \sigma_r - \sigma_\theta - 2i\tau_{r\theta} &= e^{2i\theta}\{-2(1+\hat{\nu})\overline{\zeta}F'' + 2H''\},\\ \sigma_z &= \nu(\sigma_r + \sigma_\theta),\end{aligned} \tag{7.76}$$

with Airy's function still given by the expression (7.61) with $\nu$ replaced by $\hat{\nu}$. For the problem of plane stress,

$$\begin{aligned}u_r + iu_\theta &= \frac{e^{-i\theta}}{2G}\{(3-\nu)F - (1+\nu)\zeta\overline{F}' + 2\nu(z^2-\tfrac{1}{3}h^2)\overline{F}'' + \overline{H}'\}\\ \sigma_r + \sigma_\theta + iE\omega &= 4(1+\nu)F'\\ \sigma_r - \sigma_\theta - 2i\tau_{r\theta} &= e^{2i\theta}\{-2(1+\nu)\overline{\zeta}F''+4\nu(z^2-\tfrac{1}{3}h^2)F''' + 2H''\}\\ w &= -\frac{2\nu z}{G}\operatorname{Re} F'\end{aligned} \tag{7.77}$$

with Airy's function still given by (7.62). In these solutions the complex variable itself should be in its polar form

$$\zeta = x + iy = re^{i\theta}. \tag{7.78}$$

## 7.9. Applications in Cartesian Coordinates

The close relationship between the various plane states allows us to consider, without loss of generality, only the case of generalized plane stress. We begin with a study of the contributions to stresses and displacements by polynomial functions of the variable $x + iy$. Consider first the polynomial solutions generated by the function $F$,

$$(1+\nu)F = \sum_{m=0}^{M} (a_m + ib_m)(x + iy)^m.$$

As the general result (7.39) indicates, these solutions call into play a field of areal dilations

$$\frac{\partial u}{\partial x} + \frac{\partial v}{\partial y} = \frac{1-\nu}{E}(\sigma_x + \sigma_y)$$

and a field $\omega$ of material rotations of particles, the collection being combined with shear strains so as to guarantee the existence of a displacement field. We find

$$\sigma_x = 2a_1 + 2a_2x - 6b_2y + \mathrm{Re}\sum_{m=3}^{M}[m(a_m + ib_m)\{(3-m)x^2 - (1+m)y^2 + 4ixy\}(x+iy)^{m-3}]$$

$$\sigma_y = 2a_1 + 6a_2x - 2b_2y + \mathrm{Re}\sum_{m=3}^{M}[m(a_m + ib_m)\{(1+m)x^2 + (m-3)y^2 + 4ixy\}(x+iy)^{m-3}]$$

$$\tau_{xy} = -2a_2y + 2b_2x + \mathrm{Im}\sum_{m=3}^{M}[m(m-1)(a_m+ib_m)(x^2+y^2)(x+iy)^{m-3}]$$

$$\phi = a_0x + b_0y + \mathrm{Re}\sum_{m=1}^{M}[(a_m + ib_m)(x^2 + y^2)(x+iy)^{m-1}] \tag{7.78}$$

$$2G(u+iv) = \frac{3-\nu}{1+\nu}(a_0+ib_0)+2\,\frac{1-\nu}{1+\nu}a_1(x+iy) + \frac{4}{1+\nu}b_1(-y + ix)$$
$$+ \sum_{m=2}^{M}\left[\frac{3-\nu}{1+\nu}(a_m+ib_m)(x+iy)^m - m(a_m-ib_m)(x^2+y^2)(x-iy)^{m-2}\right]. \quad (7.78)$$

The coefficients $a_0$ and $b_0$ involve only a rigid body translation and also contribute to the unproductive terms of Airy's stress function. The coefficient $b_1$ is connected with the infinitesimal rigid body rotation.

A. The state of hydrostatic stress is represented by the terms associated with the coefficient $a_1$. All directions in the plane are principal for the states of strain and stress. This situation is approximated in a thin-walled, internally pressurized, spherical reservoir with a radius of curvature approaching infinity (Figure 7.5a).

B. Uniform gradient of areal dilation. The gradient in the direction $Ox$ is associated with the terms in $a_2$; that in the direction $Oy$, with the terms in $b_2$. The displacement field already becomes relatively complex (Figures 7.5b,c).

The solutions generated by the function $H$ are simple. They are characterized by the vanishing of the rotation $\omega$ and the first invariant of the stress tensor $\sigma_x + \sigma_y$, and include only mechanisms of distortion by shearing. For

$$H' = \sum_{m=0}^{M}(p_m + iq_m)(x + iy)^m$$

one finds

$$\sigma_x - i\tau_{xy} = \sum_{m=1}^{M} m(p_m + iq_m)(x + iy)^{m-1}, \quad \sigma_y = -\sigma_x,$$

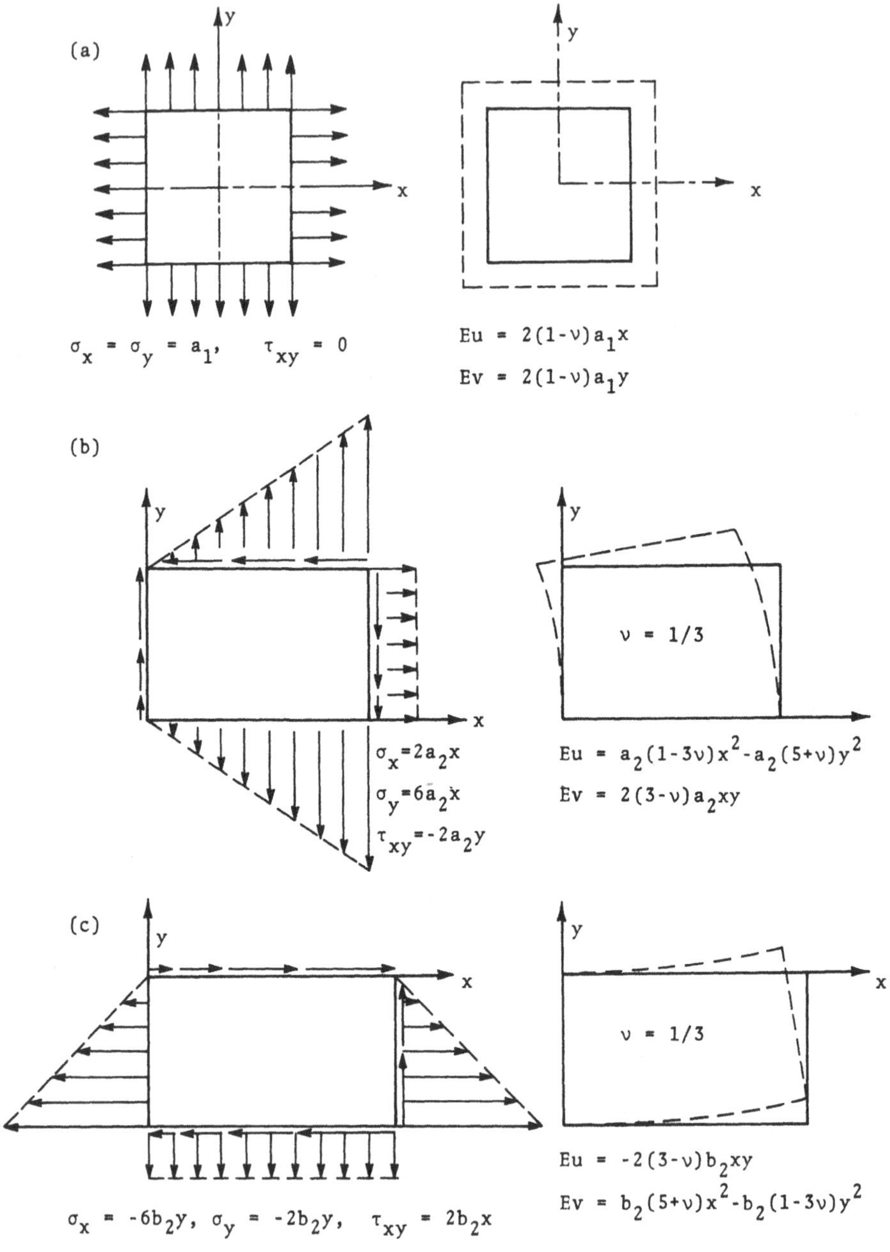
(a)
y
x
$\sigma_x = \sigma_y = a_1, \quad \tau_{xy} = 0$
$Eu = 2(1-\nu)a_1x$
$Ev = 2(1-\nu)a_1y$
(b)
y
x
$\sigma_x=2a_2x$
$\sigma_y=6a_2x$
$\tau_{xy}=-2a_2y$
$\nu = 1/3$
$Eu = a_2(1-3\nu)x^2-a_2(5+\nu)y^2$
$Ev = 2(3-\nu)a_2xy$
(c)
y
x
$\sigma_x = -6b_2y, \ \sigma_y = -2b_2y, \ \tau_{xy} = 2b_2x$
$\nu = 1/3$
$Eu = -2(3-\nu)b_2xy$
$Ev = b_2(5+\nu)x^2-b_2(1-3\nu)y^2$

Figure 7.5

$$\phi = -\text{Re}\, H = -p_{-1} - \text{Re} \sum_{m=0}^{M} \left[\frac{p_m + iq_m}{m+1} (x+iy)^{m+1}\right]$$

$$2G(u+iv) = \sum_{m=0}^{M} (p_m - iq_m)(x-iy)^m.$$

The solutions due to a coefficient $q_m$ are closely related to those of a coefficient $p_m$. If $\underset{\circ}{u}(x,y)$, $\underset{\circ}{v}(x,y)$, $\underset{\circ}{\sigma}_x(x,y)$, and $\underset{\circ}{\tau}_{xy}(x,y)$ denote the fields associated with a coefficient $p_m$, then those associated with the coefficient $q_m$ are displayed in the following table:

| | $u$ | $v$ | $\sigma_x$ | $\tau_{xy}$ |
|---|---|---|---|---|
| $m=0,4,8\ldots$ | $\underset{\circ}{v}(y,-x)$ | $-\underset{\circ}{u}(y,-x)$ | $\underset{\circ}{\sigma}_x(y,-x)$ | $\underset{\circ}{\tau}_{xy}(y,-x)$ |
| $m=1,5,9\ldots$ | $-\underset{\circ}{u}(y,-x)$ | $-\underset{\circ}{v}(y,-x)$ | $\underset{\circ}{\tau}_{xy}(y,-x)$ | $-\underset{\circ}{\sigma}_x(y,-x)$ |
| $m=2,6,10\ldots$ | $-\underset{\circ}{v}(y,-x)$ | $\underset{\circ}{u}(y,-x)$ | $-\underset{\circ}{\sigma}_x(y,-x)$ | $-\underset{\circ}{\tau}_{xy}(y,-x)$ |
| $m=3,7,11\ldots$ | $\underset{\circ}{u}(y,-x)$ | $\underset{\circ}{v}(y,-x)$ | $-\underset{\circ}{\tau}_{xy}(y,-x)$ | $\underset{\circ}{\sigma}_x(y,-x)$ |

The terms associated with $p_0$ and $q_0$ are simple translations; those with $p_{-1}$ enter the unproductive part of Airy's function.

C. Pure uniform shear. The solution associated with $p_1$ is a pure uniform shear with $Ox$ and $Oy$ as the principal axes of stress and strain. That associated with $q_1$ presents the same situation in a frame analogous, for the plane case, to the octahedral frame, i.e., one rotated 45° with respect to the first (Figures 7.6a,b).

D. Linear variation of a normal stress. The cases connected with $p_2$ and $q_2$ show the interaction between a varying shear stress on one hand, and one linearly varying component of normal stress on the other (Figure 7.6c).

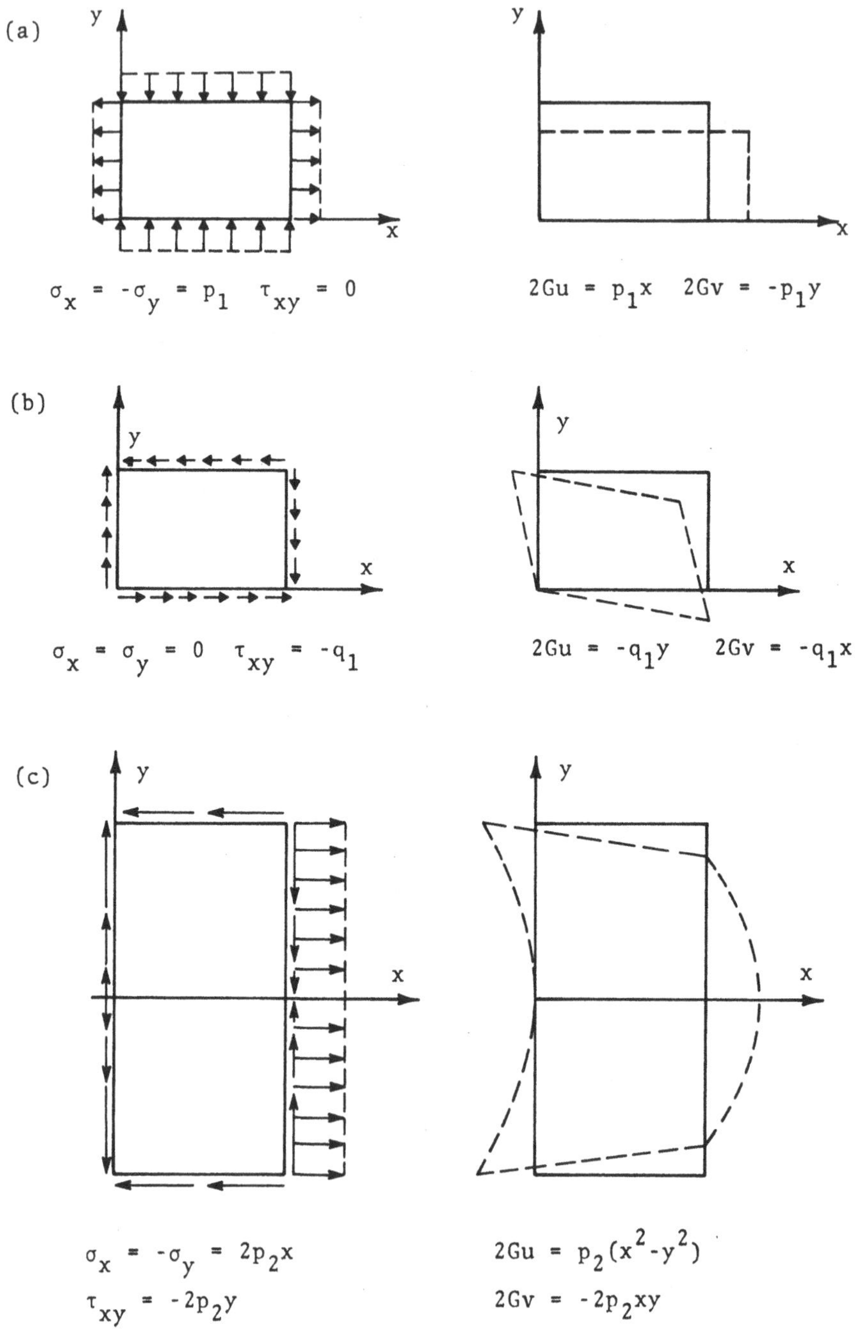
(a)
y
x
$\sigma_x = -\sigma_y = p_1 \quad \tau_{xy} = 0$
$2Gu = p_1 x \quad 2Gv = -p_1 y$
(b)
y
x
$\sigma_x = \sigma_y = 0 \quad \tau_{xy} = -q_1$
$2Gu = -q_1 y \quad 2Gv = -q_1 x$
(c)
y
x
$\sigma_x = -\sigma_y = 2p_2 x$
$\tau_{xy} = -2p_2 y$
$2Gu = p_2(x^2 - y^2)$
$2Gv = -2p_2 xy$

Figure 7.6

Simple solutions are found by combining solutions due to F and H. We seek those for which $\sigma_y \equiv 0$. This requirement is met when

$$p_1 = 2a_1, \quad p_2 = 3a_2, \quad q_2 = b_2, \quad p_3 = 0, \quad a_3 = 0, \quad q_3 = 2b_3,$$

and all the other coefficients vanish. The solution of pure shear due to $q_1$ belongs to this category, along with the following:

E. <u>Simple Extension</u>

$$\sigma_x = 4a_1, \ \sigma_y \equiv 0, \quad \tau_{xy} \equiv 0, \quad Eu = 4a_1x, \quad Ev = -4\nu a_1y.$$

F. <u>Pure Bending</u>

$$\sigma_x = -8b_2y, \quad \sigma_y \equiv 0, \quad \tau_{xy} \equiv 0, \quad Eu = -8b_2xy, \ Ev=2(1-\nu)b_2(x^2-y^2).$$

G. <u>Shear Lag</u>

$$\sigma_x = 8a_2x, \quad \tau_{xy} = -8a_2y, \quad \sigma_y \equiv 0,$$

$$Eu = 4a_2[x^2-(2+\nu)y^2], \quad Ev = -8\nu a_2xy.$$

This solution approximates the case of a longitudinal load applied by tangential tractions on the edges of a rectangular plate. It gives a uniform normal stress in a section $x =$ const., but an important warping of the cross section. When the warping is precluded, as in the neighborhood of a perfectly clamped end, the present distribution of normal stresses shows a parabolic deficiency, manifesting a phenomenon known as shear lag.

H. Bending by shear forces. If we combine the terms in $q_1$ and $b_3$ by the relation $q_1 = 12\, b_3 h^2$, then the shear stress vanishes for $y = \pm h$. The stresses become

$$\sigma_x = -24\, b_3 xy, \quad \tau_{xy} = 12\, b_3(y^2 - h^2), \quad \sigma_y \equiv 0,$$

and the associated displacements,

$$Eu = 4\, b_3 y[(2+\nu)y^2 - 3h^2] - 12\, b_3 x^2 y,$$

$$Ev = 4\, b_3 x[x^2 - 3h^2 + 3\nu y^2].$$

We again note the warping of the sections $x$ = constant. This solution clearly coincides with that of the principal bending field of the bar with rectangular section discussed in Section 6.18 H.

I. Saint-Venant's bending of a rectangular beam with flanges. The linear distribution of the normal stress in bending shows that the predominant contribution to the bending moment is from the extreme fibers. An efficient beam may therefore be formed by two concentrations of extreme fibers, the flanges, kept at a distance by a relatively thin core, which also has the very important function of transmitting the shear forces. A first approximate analysis of this type of beam amounts to considering it as a sheet (the web) in a state of plane stress subject to particular boundary conditions at $y = \pm h$. We assume that a flange, of cross-sectional area $A$, is in a state of uniform uniaxial tension or compression. If $N$ is the total force in the flange, positive in the case of tension, then

$$\frac{N}{A} = E\epsilon_x = \sigma_x - \nu\sigma_y = \frac{\partial^2\phi}{\partial y^2} - \nu\,\frac{\partial^2\phi}{\partial x^2}, \qquad y = h.$$

The second equality expresses the equality between the elongation of the flange and that of the edge of the web to which it is attached; it is a kind of compatibility condition.

There is also a condition for longitudinal equilibrium of the flange:

$$\frac{dN}{dx} = t\tau_{xy} = -t\,\frac{\partial^2\phi}{\partial x\partial y}$$

where $t$ is the thickness of the web.

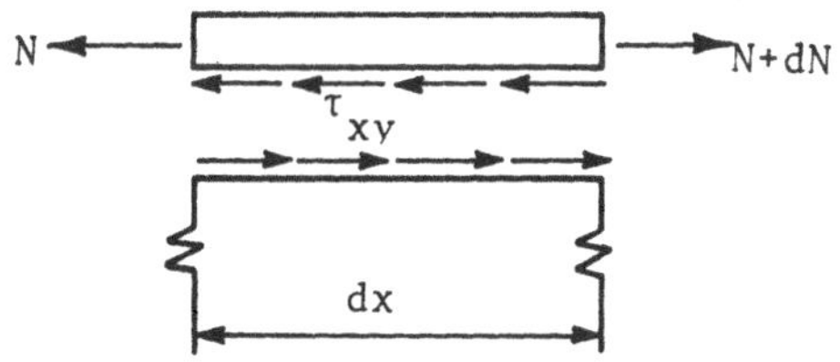


Figure 7.7

Elimination of the normal stress supplies the boundary condition for Airy's stress function:

$$\frac{d}{dx}\left[A\left(\frac{\partial^2\phi}{\partial y^2} - \nu\,\frac{\partial^2\phi}{\partial x^2}\right)\right] + t\,\frac{\partial^2\phi}{\partial x\partial y} = 0, \qquad y = h.$$

This condition can be integrated once with respect to $x$ and becomes

$$A\left(\frac{\partial^2\phi}{\partial y^2} - \nu\,\frac{\partial^2\phi}{\partial x^2}\right) + t\,\frac{\partial\phi}{\partial y} = 0, \qquad y = h,$$

which fixes the term in $\gamma y$ of the indeterminate part of Airy's function. For the bending of a symmetric beam, Airy's function is odd in $y$ and the same follows for the term in $A$ of the boundary condition. We note that for the equilibrium of the lower flange ($y = -h$) the second term should now be subtracted, assuring also the odd character of the second term.

As suggested by the case without flanges, we seek an

Airy function of the form

$$\phi = y[g(x) + \frac{y^2}{h^2} f(x)].$$

This function should be biharmonic to guarantee the compatibility of displacements:

$$(\frac{\partial^4}{\partial x^4} + 2\frac{\partial^4}{\partial y^2 \partial x^2} + \frac{\partial^4}{\partial y^4})\phi = yg^{iv} + \frac{y^3}{h^2} f^{iv} + \frac{12}{h^2} yf'' = 0.$$

In terms of the operator $D = h\frac{d}{dx}$ we have the two conditions

$$D^4 f = 0 \quad (a), \qquad D^4 g + 12D^2 f = 0 \quad (b).$$

From the boundary condition at $y = h$,

$$A(\frac{6}{h}f - \nu hg'' - \nu hf'') + tg + 3tf = 0,$$

and by introducing the web-flange parameter $m = th/A$ we obtain

$$3(2+m)f - \nu D^2 f + mg - \nu D^2 g = 0. \quad (c)$$

Applying the operator $D^2$ and using the preceding conditions (a) and (b) yields

$$3(2+m+4\nu)D^2 f + mD^2 g = 0. \quad (d)$$

Another application of $D^2$ then yields successively $D^4 g = 0$ because of (a), $D^2 f = 0$ because of (b), and $D^2 g = 0$ because of (d). Thus $f$ and $g$ are both linear functions, so (c) implies

$$3(2+m)f = -mg \quad (e)$$

and the linearity implies that

$$\sigma_y = \frac{\partial^2 \phi}{\partial x^2} \equiv 0,$$

the latter corresponding to one of Saint-Venant's assumptions in the three-dimensional problem.

The linear function $f(x)$ is connected with the bending moment; from

$$\sigma_x = \frac{\partial^2\phi}{\partial y^2} = \frac{6y}{h^2} f, \quad N = A\,\sigma_x\Big|_{y=h} = \frac{6A}{h} f,$$

we find the bending moment

$$M = 2Nh + t\int_{-h}^{h} \sigma_x y\,dy = 4A(3+m)f.$$

The distribution of the shear stress is

$$\tau_{xy} = -\frac{\partial^2\phi}{\partial x\partial y} = -g' - 3\frac{y^2}{h^2} f' = \frac{3}{h}\left[\frac{2}{m} + 1 - \frac{y^2}{h^2}\right]Df,$$

and the shear force is

$$T = t\int_{-h}^{h} \tau_{xy}dy = 4th\left(1 + \frac{3}{m}\right)f' = 4A(3+m)f' = \frac{dM}{dx}.$$

J. Transverse loading of a beam with flanges. In order to obtain a solution when the beam carries a linearly varying transverse distributed load, it is enough to add to the stress function a term in $y^5$. Calculations are carried out as before, but with

$$\phi = y\left[g(x) + \frac{y^2}{h^2} f(x) + \frac{y^4}{h^4} p(x)\right].$$

The conditions that $\phi$ be biharmonic are

$$120p + 12D^2f + D^4g = 0, \qquad \text{(a)}$$
$$40D^2p + D^4f = 0, \qquad \text{(b)}$$
$$D^4p = 0, \qquad \text{(c)}$$

whence we obtain the temporary conclusions $D^6f = 0$, $D^8g = 0$. The boundary condition caused by the flange is

$$3(2+m)f + 5(4+m)p + mg - \nu(D^2g + D^2f + D^2p) = 0. \qquad \text{(d)}$$

By applying the operator $D^6$, we find $D^6 g = 0$. Then by using $D^2$ on (a), $10D^2 p + D^4 f = 0$, and by (b) this gives $D^2 p = 0$, $D^4 f = 0$. Then, applying $D^4$ to (d), $D^4 g = 0$. Equations (b) and (c) are trivially satisfied and (a) reduces to

$$10p + D^2 f = 0. \qquad \text{(e)}$$

By use of $D^2$, (d) becomes

$$3(2+m)\, D^2 f + mD^2 g = 0. \qquad \text{(f)}$$

Replacement in (d) of $p$ and $D^2 g$ in favor of $D^2 f$ allows the final expression of $g$ as a function of $f$ and its second derivative, $f$ being a general cubic polynomial:

$$g = -3\left(\frac{2}{m} + 1\right)f + \left[\frac{1}{2} + \frac{2(1-\nu)}{m} - \frac{6\nu}{m^2}\right]D^2 f. \qquad \text{(g)}$$

Calculation of the stresses yields in particular

$$\sigma_y = \frac{\partial^2 \phi}{\partial x^2} = \frac{y}{h^2} D^2 g + \frac{y^3}{h^4} D^2 f = \left[\frac{y^3}{h^4} - 3\left(\frac{2}{m} + 1\right)\frac{y}{h^2}\right] D^2 f$$

and the transverse load applied to the upper flange ($y = h$) is

$$p_y = t\, \sigma_y \Big|_{y=h} = -\frac{t}{h}\left(2 + \frac{6}{m}\right) D^2 f.$$

The same load is obviously applied to the lower flange ($y = -h$). This load may vary linearly with $x$, and its determination defines the coefficients of the polynomial $D^2 f$ of the first degree. For example, if this load were zero, we would return to the case in the previous section. In general,

$$\sigma_x = \frac{\partial^2 \phi}{\partial y^2} = \frac{6}{h^2} yf + \frac{20}{h^4} y^3 p = \frac{6}{h^2} yf - \frac{2}{h^4} y^3 D^2 f$$

and the distributed load induces a distribution of normal stress which is no longer proportional to the distance from the neutral fiber. With

$$N = A(\sigma_x - \nu\sigma_y)\Big|_{y=h} = \frac{2A}{h}\,[3f - \{1 - \nu(1+ \tfrac{3}{m})\}D^2 f]$$

it is possible to express the bending moment as

$$M = 2Nh + t\int_{-h}^{h} \sigma_x y\,dy = 4ht(f - \tfrac{1}{5}D^2 f) + 4A[3f-\{1-(1+ \tfrac{3}{m})\}D^2 f]$$

$$= 4A(3+m)f - 4A[1+ \tfrac{m}{5} - \nu(1+ \tfrac{3}{m})]D^2 f.$$

Thus, if $D^2 f$ is known from the applied load, knowledge of the bending moment determines the function $f$ completely. For the shear stress and shear force we have

$$\tau_{xy} = -\frac{\partial^2\phi}{\partial x\partial y} = -g' - 3\,\frac{y^2}{h^2}\,f' - 5\,\frac{y^4}{h^4}\,p'$$

$$= \frac{3}{h}[\frac{2}{m} + 1 - \frac{y^2}{h^2}]Df - \frac{1}{2h}[1+ \frac{4(1-\nu)}{m} - \frac{12\nu}{m^2} - \frac{y^4}{h^4}]D^3 f,$$

$$T = t\int_{-h}^{h} \tau_{xy}\,dy = 4t(\frac{3}{m} + 1)Df - 4t[\frac{1}{5} + \frac{1}{m} - \frac{\nu}{m}(1+ \frac{3}{m})D^3 f] = \frac{dM}{dx}$$

Finally,

$$\frac{dT}{dx} = \frac{4t}{h}\,(\frac{3}{m} + 1)D^2 f = -2p_y.$$

This calculation is a control intended to verify the exactness of the equations of global equilibrium of the beam, which are a necessary consequence of the particular equilibrium of each element. The case in which the transverse load is applied to a single flange may be handled by superposing on the preceding solution a solution involving opposing loads and depending on an Airy's function developed in even powers of $y$.

## 7.10. Applications in Polar Coordinates

There is an extensive field of applications for the general solution of the plane state for a region containing a circular ring centered at the origin. We require that the stresses be univalent there, which implies that the functions $H''$, $F''$, and $\mathrm{Re}[F']$ are univalent in the ring. Laurent's theorem then assures us of the existence of expansions for $H''$ and $F''$, converging in the ring, of the form

$$\sum_{m=-\infty}^{-\infty} (p_m + iq_m)\zeta^m .$$

After a double integration, we write

$$H = (\gamma+i\delta)\zeta(\ln\zeta-1) + (\alpha+i\beta)\ln \zeta + \sum_{-\infty}^{\infty}(\alpha_m+i\beta_m)\zeta^m,$$

$$(1+\nu)F = \eta\zeta(\ln\zeta-1) + (\lambda+i\mu)\ln \zeta + \sum_{-\infty}^{\infty}(\lambda_m+i\mu_m)\zeta^m .$$

If the stresses are to be univalent, so must $\mathrm{Re}[F']$, and this requires that $\eta$ be real.

With the substitutions

$$\zeta = re^{i\theta}, \qquad \ln \zeta = \ln r + i\theta$$

we obtain the expansions in polar coordinates of the displacements and the stresses:

$$\begin{aligned} 2Gu_r = & -\theta\cos\theta\left(\delta + \frac{3-\nu}{1+\nu}\mu\right)+\theta\sin\theta\left(-\gamma + \frac{3-\nu}{1+\nu}\lambda\right) \\ & + \eta r\left(2\,\frac{1-\nu}{1+\nu}\ln r - \frac{3-\nu}{1+\nu}\right) + 2\,\frac{1-\nu}{1+\nu}\lambda_1 r + \frac{\alpha}{r} \\ & + \cos\theta\left[\alpha_1-\lambda+\gamma\ln r - \frac{\alpha_{-1}}{r^2} + \frac{1-3\nu}{1+\nu}\lambda_2 r^2 + \frac{3-\nu}{1+\nu}(\lambda_0+\lambda\ln r)\right] \\ & + \sin\theta\left[-\beta_1-\mu-\delta\ln r - \frac{\beta_{-1}}{r^2} - \frac{1-3\nu}{1+\nu}\mu_2 r^2 + \frac{3-\nu}{1+\nu}(\mu_0+\mu\ln r)\right] \end{aligned}$$

$$+\sum_{m=2}^{\infty}\cos m\theta\left[\frac{2-m-\nu(2+m)}{1+\nu}\lambda_{m+1}r^{m+1}+m\alpha_m r^{m-1}+\frac{2+m-\nu(2-m)}{1+\nu}\frac{\lambda_{1-m}}{r^{m-1}}-m\frac{\alpha_{-m}}{r^{m+1}}\right]$$

$$+\sum_{m=2}^{\infty}\sin m\theta\left[-\frac{2-m-\nu(2+m)}{1+\nu}\mu_{m+1}r^{m+1}-m\beta_m r^{m-1}+\frac{2+m-\nu(2-m)}{1+\nu}\frac{\mu_{1-m}}{r^{m-1}}+m\frac{\beta_{-m}}{r^{m+1}}\right]$$

(7.78')

$$2Gu_\theta = \frac{4}{1+\nu}\eta\theta r+\theta\cos\theta\left(-\gamma+\frac{3-\nu}{1+\nu}\lambda\right)+\theta\sin\theta\left(\delta+\frac{3-\nu}{1+\nu}\mu\right)+\frac{4}{1+\nu}\mu_1 r-\frac{\beta}{r}$$

$$+\cos\theta\left[-\beta_1+\mu-\delta\ell nr+\frac{\beta_{-1}}{r^2}+\frac{5+\nu}{1+\nu}\mu_2 r^2+\frac{3-\nu}{1+\nu}(\mu_o+\mu\ell nr)\right]$$

$$+\sin\theta\left[-\alpha_1-\lambda-\gamma\ell nr-\frac{\alpha_{-1}}{r^2}+\frac{5+\nu}{1+\nu}\lambda_2 r^2-\frac{3-\nu}{1+\nu}(\lambda_o+\lambda\ell nr)\right]$$

$$+\sum_{m=2}^{\infty}\cos m\theta\left[\frac{4+m+\nu m}{1+\nu}\mu_{m+1}r^{m+1}-m\beta_m r^{m-1}+\frac{3-\nu}{1+\nu}\frac{\mu_{1-m}}{r^{m-1}}+m\frac{\beta_{-m}}{r^{m+1}}\right]$$

$$+\sum_{m=2}^{\infty}\sin m\theta\left[\frac{4+m+\nu m}{1+\nu}\lambda_{m+1}r^{m+1}-m\alpha_m r^{m-1}-\frac{3-\nu}{1+\nu}\frac{\lambda_{1-m}}{r^{m-1}}-m\frac{\alpha_{-m}}{r^{m+1}}\right]$$

(7.78")

$$\phi = \beta\theta+\theta\cos\theta(\delta-\mu)r+\theta\sin\theta(\lambda+\gamma)r$$

$$+\eta r^2(\ell nr-1)+\lambda_1 r^2-\alpha\ell nr-\alpha_o$$

$$+\cos\theta\left[\lambda r\ell nr+\lambda_o r+\lambda_2 r^3-\gamma r(\ell nr-1)-\frac{\alpha_{-1}}{r}-\alpha_1 r\right]$$

$$+\sin\theta\left[\mu r\ell nr+\mu_o r-\mu_2 r^3+\delta r(\ell nr-1)-\frac{\beta_{-1}}{r}+\beta_1 r\right]$$

$$+\sum_{m=2}^{\infty}\cos m\theta\left[\lambda_{m+1}r^{m+2}-\alpha_m r^m+\frac{\lambda_{1-m}}{r^{m-2}}-\frac{\alpha_{-m}}{r^m}\right]$$

$$+\sum_{m=2}^{\infty}\sin m\theta\left[-\mu_{m+1}r^{m+2}+\beta_m r^m+\frac{\mu_{1-m}}{r^{m-2}}-\frac{\beta_{-m}}{r^m}\right] \qquad (7.79)$$

$$\sigma_r = \eta(2\ln r-1) + 2\lambda_1 - \frac{\alpha}{r^2}$$
$$+ \cos\theta\left(\frac{3\lambda+\gamma}{r} + 2\lambda_2 r + \frac{2\alpha_{-1}}{r^3}\right)$$
$$+ \sin\theta\left(\frac{3\mu-\delta}{r} - 2\mu_2 r + \frac{2\beta_{-1}}{r^3}\right)$$
$$+ \sum_{m=2}^{\infty} \cos m\theta\Big[-(m-2)(m+1)\lambda_{m+1}r^m + m(m-1)\alpha_m r^{m-2} - (m-1)(m+2)\frac{\lambda_{1-m}}{r^m}$$
$$+ m(m+1)\frac{\alpha_{-m}}{r^{m+2}}\Big]$$
$$+ \sum_{m=2}^{\infty} \sin m\theta\Big[(m-2)(m+1)\mu_{m+1}r^m - m(m-1)\beta_m r^{m-2} - (m-1)(m+2)\frac{\mu_{1-m}}{r^m}$$
$$+ m(m+1)\frac{\beta_{-m}}{r^{m+2}}\Big] \qquad (7.80)$$

$$\tau_{r\theta} = \frac{\beta}{r^2} + \cos\theta\left(2\mu_2 r - \frac{\delta+\mu}{r} - 2\frac{\beta_{-1}}{r^3}\right) + \sin\theta\left(2\lambda_2 r + \frac{\lambda-\gamma}{r} + 2\frac{\alpha_{-1}}{r^3}\right)$$
$$+ \sum_{m=2}^{\infty} \cos m\theta\Big[m(m+1)\Big(\mu_{m+1}r^m - \frac{\beta_{-m}}{r^{m+2}}\Big) + m(m-1)\Big(\frac{\mu_{1-m}}{r^m} - \beta_m r^{m-2}\Big)\Big]$$
$$+ \sum_{m=2}^{\infty} \sin m\theta\Big[m(m+1)\Big(\lambda_{m+1}r^m + \frac{\alpha_{-m}}{r^{m+2}}\Big) - m(m-1)\Big(\alpha_m r^{m-2} + \frac{\lambda_{1-m}}{r^m}\Big)\Big]$$
$$(7.81)$$

$$\sigma_\theta = \eta(2\ln r+1) + 2\lambda_1 + \frac{\alpha}{r^2}$$
$$+ \cos\theta\left(6\lambda_2 r + \frac{\lambda-\gamma}{r} - 2\frac{\alpha_{-1}}{r^3}\right) + \sin\theta\left(-6\mu_2 r + \frac{\mu+\delta}{r} - 2\frac{\beta_{-1}}{r^3}\right)$$
$$+ \sum_{m=2}^{\infty} \cos m\theta\Big[(m+1)(m+2)\lambda_{m+1}r^m - m(m-1)\alpha_m r^{m-2} + (m-1)(m-2)\frac{\lambda_{1-m}}{r^m}$$
$$- m(m+1)\frac{\alpha_{-m}}{r^{m+2}}\Big]$$
$$+ \sum_{m=2}^{\infty} \sin m\theta\Big[-(m+1)(m+2)\mu_{m+1}r^m + m(m-1)\beta_m r^{m-2} + (m-1)(m-2)\frac{\mu_{1-m}}{r^m}$$
$$- m(m+1)\frac{\beta_{-m}}{r^{m+2}}\Big]. \qquad (7.82)$$

The first line of the expressions for the displacements gives the multivalent terms. The difference between the values for $\theta + 2\pi$ and $\theta$ are

$$2G\Delta u_r = -2\pi\cos\theta\,(\delta + \frac{3-\nu}{1+\nu}\mu) + 2\pi\sin\theta\,(-\gamma + \frac{3-\nu}{1+\nu}\lambda)$$

$$2G\Delta u_\theta = \frac{8\pi}{1+\nu}\eta r + 2\pi\cos\theta\,(-\gamma + \frac{3-\nu}{1+\nu}\lambda) + 2\pi\sin\theta\,(\delta + \frac{3-\nu}{1+\nu}\mu).$$

In Volterra's theory of dislocations, if the origin is to be a center of reduction, i.e., the edge of a half-plane suffering constant displacement and rotation discontinuity, then we must have

$$\Delta u = \Delta p - y\Delta\omega, \quad \Delta v = \Delta q + x\Delta\omega,$$

where $(\Delta p, \Delta q, \Delta\omega)$ are the constant values of discontinuity in $(u, v, \omega)$. In polar coordinates this becomes

$$\Delta u_r = \Delta p\cos\theta + \Delta q\sin\theta$$

$$\Delta u_\theta = -\Delta p\sin\theta + \Delta q\cos\theta + r\Delta\omega.$$

By comparison, we have*

$$\Delta\omega = \frac{8\pi}{E}\eta, \quad \Delta p = -\frac{2\pi}{E}[(1+\nu)\delta+(3-\nu)\mu],$$

$$\Delta q = \frac{2\pi}{E}[-(1+\nu)\gamma+(3-\nu)\lambda].$$

The rigid body displacement modes are represented by the coefficients $(\alpha_1, \beta_1, \mu_1)$. As we have seen, the multivalency of Airy's function is connected with the resultant of the stresses applied to the interior of the cavity. If it

*Editor's note: There are useful examples where one or more of the constants $(\eta,\delta,\mu,\delta,\lambda)$ does not vanish but the displacements are univalent *in the range of interest*. For example, when the region subtends less than a full circular arc, the solution outside the interval, say, $-\pi+\varepsilon < \theta < \pi-\varepsilon$ is of no concern. See, for example, section 7.10C.

vanishes, $\beta = 0$, $\delta = \mu$, and $\gamma = -\lambda$. The unproductive terms of Airy's function are those connected with the coefficients $(\alpha_o, \lambda_o, \mu_o)$.

The foregoing method, based essentially on the complex representation of *Muskhelishvili-Kolosov*, has the advantage of yielding the displacements along with the stresses. If only the latter are desired, their Fourier expansion can be found by the following more direct approach. Let $a_r$ and $a_\theta$ denote, respectively, the radial and tangential components of the gradient of Airy's function:

$$a_r = \frac{\partial\phi}{\partial r}, \quad a_\theta = \frac{1}{r}\frac{\partial\phi}{\partial\theta}.$$

Then the Laplacian of Airy's function will be

$$\nabla^2\phi = \text{div grad } \phi = \frac{1}{r}\left[\frac{\partial}{\partial r}(ra_r) + \frac{\partial a_\theta}{\partial\theta}\right]$$

or, for Laplace's operator in polar coordinates,

$$\nabla^2 = \frac{1}{r}\frac{\partial}{\partial r}(r\frac{\partial}{\partial r}) + \frac{1}{r^2}\frac{\partial^2}{\partial\theta^2}. \tag{7.83}$$

A harmonic function of the form $f(r) \cos m\theta$ or $f(r) \sin m\theta$ thus has its radial function $f(r)$ controlled by the differential equation

$$\frac{1}{r}\frac{d}{dr}(r\frac{df}{dr}) - \frac{m^2}{r^2} f = 0.$$

The change of variable $r = e^t$ transforms this to

$$\frac{d^2f}{dt^2} - m^2f = 0.$$

For the general solution when $m \neq 0$ we thus get

$$f = Ae^{mt} + Be^{-mt} = Ar^m + Br^{-m}. \tag{7.84}$$

We also know, from section 7.2, that the product of a

harmonic function and $r^2 = x^2 + y^2$ yields a biharmonic function. Accordingly, if

$$g(r) = Ar^m + Br^{-m} + r^2(Cr^m + Dr^{-m}) \tag{7.85}$$

then $g(r) \cos m\theta$ and $g(r) \sin m\theta$ are biharmonic. Because for $m \geq 2$ the function $g(r)$ contains four independent constants, we have the *general* solution except for the cases $m = 0$ and $m = 1$.

For $m = 0$ the general solution in the harmonic case is

$$f = A + Bt = A + B \ln r, \tag{7.86}$$

and a biharmonic function depending only on $r$ thus has the general expression

$$g(r) = A + B \ln r + r^2(C + D \ln r). \tag{7.87}$$

For $m = 1$ the harmonic solution (7.84) is general. Now recall from section (7.2) that a biharmonic function results upon multiplying a harmonic function by $x = r \cos \theta$ or $y = r \sin \theta$. Thus we can find the general expression of a biharmonic function of the type $g(r) \cos \theta$ or $g(r) \sin \theta$ by adding to (7.85) the product of $r$ itself with a harmonic function of $r$ alone, such as (7.86). For $m = 1$, the term in $D$ in (7.85) has the same form as the term in $A$ in (7.86), and will therefore be replaced by a term in $r \ln r$:

$$g(r) = Ar + Br^{-1} + Cr^3 + Dr \ln r. \tag{7.88}$$

These results confirm the structure obtained in (7.79) for the univalent part of Airy's function. We now consider several examples.

A. Circular aperture with traction-free circumference in a plate in plane stress. Let $r = a$ be the boundary of the aperture on which the stresses $\sigma_r$ and $\tau_{r\theta}$ should vanish. This property may be expressed rather simply by using Airy's function. Indeed, we know that it is always possible to annul this function and its first derivatives at an arbitrary point by a judicious choice of the unproductive parameters. If we choose a point on the circle $r = a$, Airy's function and its first derivatives remain zero because of the properties (7.55), since we have $X_p = Y_p = M_p = 0$ everywhere in the absence of surface tractions on the boundary of the aperture. We may therefore replace the conditions

$$\sigma_r = 0, \quad \tau_{r\theta} = 0 \quad \text{at} \quad r = a$$

by the equivalent conditions

$$\phi = 0, \quad \partial\phi/\partial r = 0 \quad \text{at} \quad r = a.$$

These lead to the relations $\beta = 0$, $\mu = \delta$, $\lambda = -\gamma$, which make $\phi$ univalent as was anticipated, and to the relations

$$2\alpha_o = -a^2\eta + (1-2\ell na)\alpha, \qquad 2\lambda_1 = \frac{\alpha}{a^2} + (1-2\ell na)\eta,$$

$$\alpha_1 = 2a^2\lambda_2 + \lambda_o - 2\gamma\ell na, \qquad \alpha_{-1} = a^2\gamma - \lambda_2 a^4,$$

$$\beta_1 = -\mu_o + 2a^2\mu_2 - 2\delta\ell na, \qquad \beta_{-1} = -a^2\delta + \mu_2 a^4,$$

$$\left.\begin{aligned} \alpha_{-m} &= -m\, a^{2m+2}\lambda_{m+1} + (m-1)a^{2m}\alpha_m, \\ \lambda_{1-m} &= -(m+1)a^{2m}\lambda_{m+1} + ma^{2m-2}\alpha_m, \\ \beta_{-m} &= ma^{2m+2}\mu_{m+1} - (m-1)\, a^{2m}\beta_m, \\ \mu_{1-m} &= (m+1)a^{2m}\mu_{m+1} - ma^{2m-2}\beta_m. \end{aligned}\right\} \quad m \geq 2$$

With the notation $\rho = r/a$ this yields for Airy's function the formula

$$\begin{aligned}\phi &= \eta a^2(\rho^2 \ell n\rho + \frac{1-\rho^2}{2}) - \alpha(\ell n\rho + \frac{1-\rho^2}{2}) \\ &+ (\rho - \frac{1}{\rho} - 2\rho\ell n\rho)(\gamma\cos\theta - \delta\sin\theta)a + \frac{(\rho^2-1)^2}{\rho}(\lambda_2\cos\theta - \mu_2\sin\theta)a^3 \\ &+ \sum_{m=2}^{\infty}[\rho^{m+2} - (m+1)\rho^{2-m} + m\rho^{-m}](\lambda_{m+1}\cos m\theta - \mu_{m+1}\sin m\theta)a^{m+2} \\ &+ \sum_{m=2}^{\infty}[\rho^m - m\rho^{2-m} + (m-1)\rho^{-m}]\ (\alpha_m\cos m\theta - \beta_m\sin m\theta)a^m. \qquad (7.89)\end{aligned}$$

Now suppose that the plate extends to infinity in all directions. We must then suppress in Airy's function the terms contributing stresses which tend to infinity with $\rho$. Only the coefficients $(\alpha, \gamma, \delta, \alpha_2, \beta_2)$ remain, and the stresses become

$$\begin{aligned}a^2\sigma_\theta &= \frac{\partial^2\phi}{\partial\rho^2} = \alpha(1 + \frac{1}{\rho^2}) - 2(\frac{1}{\rho} + \frac{1}{\rho^3})(\gamma\cos\theta - \delta\sin\theta) \\ &\qquad - 2(1 + \frac{3}{\rho^4})a^2(\alpha_2\cos 2\theta - \beta_2\sin 2\theta), \\ a^2\tau_{r\theta} &= -\frac{\partial}{\partial\rho}(\frac{1}{\rho}\frac{\partial\phi}{\partial\theta}) = 2(\frac{1}{\rho^3} - \frac{1}{\rho})(\gamma\sin\theta + \delta\cos\theta) \\ &\qquad - 2(1 + \frac{2}{\rho^2} - \frac{3}{\rho^4})a^2(\alpha_2\sin 2\theta + \beta_2\cos 2\theta), \\ a^2\sigma_r &= \frac{1}{\rho^2}\frac{\partial^2\phi}{\partial\theta^2} + \frac{1}{\rho}\frac{\partial\phi}{\partial\rho} = \alpha(1 - \frac{1}{\rho^2}) - 2(\frac{1}{\rho} - \frac{1}{\rho^3})(\gamma\cos\theta - \delta\sin\theta) \\ &\qquad + 2(1 - \frac{4}{\rho^2} + \frac{3}{\rho^4})a^2(\alpha_2\cos 2\theta - \beta_2\sin 2\theta).\end{aligned}$$

The terms in $\gamma$ and $\delta$ do not contribute to the state of stress prevailing at a great distance, but they decrease rather slowly, namely, as $\rho^{-1}$. These terms are connected with multivalent displacements and therefore represent regular states of self-stress which may arise from the closing of dislocations of the Weingarten-Volterra type. (The self-stress arising from the closing of a dislocation in $\Delta\omega$ corresponds to the terms in $\eta$; these have been dropped because $\sigma_r$ and $\sigma_\theta$ behave at great distance as $\ln \rho$.)

The stresses due to the remaining terms have significant interpretations at great distance in polar or Cartesian coordinates. For this purpose one uses the relations

$$\sigma_\theta + \sigma_r = \sigma_x + \sigma_y = 2a^{-2}\alpha,$$

$$\sigma_x - \sigma_y - 2i\tau_{xy} = (\sigma_r - \sigma_\theta - 2i\tau_{r\theta})e^{-2i\theta} = 4(\alpha_2 + i\beta_2).$$

In the *hydrostatic* case, where $\alpha_2 = \beta_2 = 0$, the stresses at great distance are $\sigma_x = \sigma_y = \sigma_r = \sigma_\theta = \alpha a^{-2} = \sigma_\infty$, while at finite distance

$$\sigma_\theta = \sigma_\infty(1 + \rho^{-2}), \quad \sigma_r = \sigma_\infty(1 - \rho^{-2}), \quad \tau_{r\theta} \equiv 0.$$

The perturbation of the state of hydrostatic stress decreases toward infinity as $\rho^{-2}$. The maximum principal stress occurs at the edge of the aperture where $\sigma_\theta = 2\sigma_\infty$.

In the case of *pure shear*, where $\alpha = \alpha_2 = 0$, the stresses at great distance are $\tau_{xy} = -2\beta_2 = \tau_\infty$, $\sigma_x = \sigma_y = 0$, while at finite distance

$$\sigma_\theta = -\tau_\infty(1 + 3\rho^{-4})\sin 2\theta, \quad \sigma_r = \tau_\infty(1 - 4\rho^{-2} + 3\rho^{-4})\sin 2\theta,$$

$$\tau_{r\theta} = \tau_\infty(1 + 2\rho^{-2} - 3\rho^{-4}) \cos 2\theta.$$

The perturbation of $\sigma_\theta$ decreases more rapidly than the others but it is responsible for a local maximum $\sigma_\theta = \pm 4\tau_\infty$ of one principal stress at the edge of the aperture.

For the case of *pure tension*, we require, at great

distance, $\sigma_x = \sigma_\infty$, $\sigma_y = \tau_{xy} = 0$; accordingly we choose $\beta_2 = 0$ and $\sigma_\infty = 4\alpha_2 = 2a^{-2}\alpha$. The stresses at finite distance are

$$\frac{\sigma_\theta}{\sigma_\infty} = \frac{1}{2}\,(1 + \frac{1}{\rho^2}) - \frac{1}{2}(1 + \frac{3}{\rho^4}) \cos 2\theta,$$

$$\frac{\tau_{r\theta}}{\sigma_\infty} = -\frac{1}{2}(1 + \frac{2}{\rho^2} - \frac{3}{\rho^4}) \sin 2\theta,$$

$$\frac{\sigma_r}{\sigma_\infty} = \frac{1}{2}(1 - \frac{1}{\rho^2}) + \frac{1}{2}(1 - \frac{4}{\rho^2} + \frac{3}{\rho^4}) \cos 2\theta.$$

The maximum local principal stress is $\sigma_\theta = 3\sigma_\infty$, occurring at $\theta = \pm\pi/2$.

The question arises of recognizing the level at which the presence of the aperture may involve a danger of exceeding the elastic limit. Using Tresca's criterion, one is led to compare the local maximum shear stress with that prevailing at a great distance. Their ratio is the *stress concentration factor*. It should be observed that, in the direction perpendicular to the slab, $\sigma_z = 0$ is one of the principal stresses. In the hydrostatic case, the maximum shear stress at great distance is $\sigma_\infty/2$, whereas locally at $\rho = 1$, it is $\sigma_\infty$. The stress concentration factor is 2.

In the case of pure shear, at great distance the maximum shear stress is $\tau_\infty$; locally, $2\tau_\infty$. The stress concentration factor is 2.

In the case of pure tension, at great distance we have $\sigma_\infty/2$; locally, $3\sigma_\infty/2$. Here the stress concentration factor is 3.

Slightly different results follow from the criterion of Hüber-Hencky-von Mises. According to (5.58) the criterion may be written

$$(\sigma_x-\sigma_y)^2+(\sigma_y-\sigma_z)^2+(\sigma_z-\sigma_x)^2+6(\tau_{xy}^2 + \tau_{yz}^2 + \tau_{zx}^2) < 2\sigma_e^2.$$

In the plane state of stress, this reduces to

$$\sigma_x^2 + \sigma_y^2 - \sigma_x\sigma_y + 3\tau_{xy}^2 < \sigma_e^2.$$

In the form

$$\tfrac{3}{4}(\sigma_x-\sigma_y-2i\tau_{xy})(\sigma_x-\sigma_y+2i\tau_{xy}) + \tfrac{1}{4}(\sigma_x+\sigma_y)^2 < \sigma_e^2$$

this result yields immediately the expression in polar coordinates

$$\sigma_r^2 + \sigma_\theta^2 - \sigma_r\sigma_\theta + 3\tau_{r\theta}^2 < \sigma_e^2$$

which is another confirmation of the fact that the left hand side is a Cartesian invariant. One may therefore calculate its local maximum analytically. The square root of its ratio with its value at infinity will be the stress concentration factor. The local maxima are found to occur at the same positions as in the previous discussion.

For hydrostatic stress, at great distance we have $\sigma_\infty^2 + \sigma_\infty^2 - \sigma_\infty^2 = \sigma_\infty^2$; locally $4\sigma_\infty^2$; the stress concentration factor is 2. For pure shear, at great distance $3\tau_\infty^2$; locally $16\,\tau_\infty^2$; the factor is $4/\sqrt{3} = 2.3094$. For pure tension, at great distance $\sigma_\infty^2$; locally $9\sigma_\infty^2$; the factor is 3.

B. Volterra's dislocation of the circular ring. Consider a circular ring of thickness $t$ bounded by the circles $r = a$ and $r = b$ and free of applied loads. We may begin with Airy's function in the form (7.89), which already guarantees the vanishing of $\sigma_r$ and $\tau_{r\theta}$ when $r = a$. To obtain the same vanishing when $r = b$, add the unproductive form $m + \rho(n \cos\theta + p \sin\theta)$ so as to allow $\phi$ and its first derivatives to vanish instead at $r = b$ or $\rho = b/a \equiv \beta$. With $\phi$ given by (7.89), the conditions

$$\phi + m + \rho(n \cos \theta + p \sin \theta) = 0,$$

$$\frac{\partial\phi}{\partial\rho} + n \cos \theta + p \sin \theta = 0 \quad \text{at} \quad \rho = \beta$$

furnish new relations among the remaining coefficients. The annihilation of the term independent of $\theta$ in the first equation merely determines the unproductive coefficient $m$ and is of no interest. In the second, however, it yields

$$2\eta a^2\beta^2 \ln \beta = \alpha(1 - \beta^2).$$

For the terms in $\sin \theta$ and $\cos \theta$ we eliminate the unproductive coefficients $n$ and $p$ between the two new conditions, getting

$$\gamma = (1+\beta^2)a^2\lambda_2, \qquad \delta = (1+\beta^2)a^2\mu_2.$$

For $m \geq 2$ the coefficients of $\sin m\theta$ and $\cos m\theta$ vanish because they satisfy linear homogeneous equations with nonsingular determinant. Omitting the unproductive terms, which served merely as auxiliary in the calculations, Airy's function (7.89) becomes

$$\phi = -\eta a^2 P(\rho) - a(\gamma \cos \theta - \delta \sin \theta)F(\rho) \tag{7.90}$$

where

$$P(\rho) = -\rho^2 \ln \rho + \frac{1}{2}(\rho^2-1) + \frac{2\beta^2 \ln \beta}{\beta^2-1}\left(\frac{\rho^2-1}{2} - \ln \rho\right), \tag{7.91}$$

$$F(\rho) = \frac{1}{\rho} - \rho + 2\rho \ln \rho - \frac{1}{\beta^2+1}\,\frac{(\rho^2-1)^2}{\rho}. \tag{7.92}$$

From (7.71), the stresses are

$$\sigma_\theta = 2\eta[1+\ln\frac{r}{a} - \frac{b^2}{b^2-a^2}(1+\frac{a^2}{r^2})\ln\frac{b}{a}] + [\frac{6r}{a^2+b^2} - \frac{2}{r} - \frac{2a^2b^2}{(a^2+b^2)r^3}]\cdot$$
$$\cdot(\gamma\cos\theta - \delta\sin\theta),$$

$$\tau_{r\theta} = \frac{2(r^2-a^2)(r^2-b^2)}{(a^2+b^2)r^3}(\gamma\sin\theta + \delta\cos\theta),$$

$$\sigma_r = 2\eta[\ln\frac{r}{a} + \frac{b^2}{b^2-a^2}(\frac{a^2}{r^2}-1)\ln\frac{b}{a}] + \frac{2(r^2-a^2)(r^2-b^2)}{(a^2+b^2)r^3}(\gamma\cos\theta-\delta\sin\theta),$$

and from (7.78'), (7.78"), and (7.23), the displacements and rotation are

$$2Gu_r = \frac{2(1-\nu)}{1+\nu}\eta[r\ln\frac{r}{a} + r(\frac{1}{2} - \frac{b^2}{b^2-a^2}\ln\frac{b}{a})] - \eta[\frac{3-\nu}{1+\nu}r + (\frac{2a^2b^2}{b^2-a^2}\ln\frac{b}{a})\frac{1}{r}]$$
$$+(\gamma\cos\theta-\delta\sin\theta)[1+\frac{1-3\nu}{1+\nu}\frac{r^2}{b^2+a^2} - \frac{2(1-\nu)}{1+\nu}\ln\frac{r}{a} - \frac{a^2b^2}{b^2+a^2}\frac{1}{r^2}]$$
$$-\frac{4}{1+\nu}(\gamma\sin\theta + \delta\cos\theta)\theta + \alpha_1\cos\theta - \beta_1\sin\theta,$$

$$2Gu_\theta = \frac{4}{1+\nu}\eta r\theta + (\gamma\sin\theta+\delta\cos\theta)[1+\frac{2(1-\nu)}{1+\nu}\ln\frac{r}{a} - \frac{a^2b^2}{b^2+a^2}\frac{1}{r^2} +$$
$$+\frac{5+\nu}{1+\nu}\frac{r^2}{b^2+a^2}]$$
$$-\frac{4}{1+\nu}(\gamma\cos\theta-\delta\sin\theta)\theta + \frac{4}{1+\nu}\mu_1 r - (\alpha_1\sin\theta + \beta_1\cos\theta),$$

$$E\omega = 4\eta\theta + 4(\frac{1}{r} + \frac{2r}{a^2+b^2})(\gamma\sin\theta + \delta\cos\theta).$$

The terms in $(\alpha_1, \beta_1, \mu_1)$ represent only rigid modes. The parameters $(\lambda, \gamma, \delta)$, with which on the other hand the stresses and strains vary, may be connected with the dislocations in an arbitrarily chosen section $\theta$ = const. Between $\theta = 0$ and $\theta = 2\pi$, for example,

$$E\Delta\omega = 8\pi\eta, \quad E\Delta u_r = -8\pi\delta, \quad E\Delta u_\theta = -8\pi\gamma + E\Delta\omega r.$$

One may also connect them with the resultant of the stresses in the section. Direct integration yields

$$N_\theta \equiv t \int_a^b \sigma_\theta \, dr = -8\gamma S$$

$$N_r \equiv t \int_a^b \tau_{r\theta} \, dr = -8\delta S$$

$$\mu \equiv t \int_a^b \sigma_\theta \, r \, dr = 4B\eta$$

where

$$S = \frac{t}{4}\left(\ln \frac{b}{a} - \frac{b^2-a^2}{b^2+a^2}\right), \tag{7.93}$$

$$B = \frac{t}{4}\left[\frac{b^2-a^2}{2} - \frac{2a^2b^2}{b^2-a^2}\left(\ln \frac{b}{a}\right)^2\right]. \tag{7.94}$$

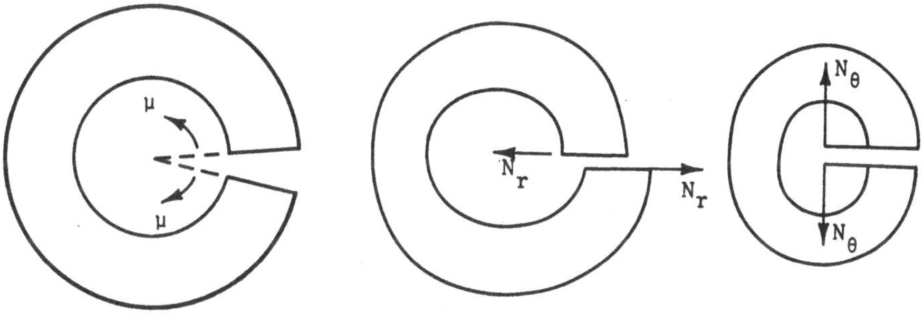


(a) $\delta\omega < 0$, $\mu < 0$. (b) $\Delta u_r > 0$, $N_r > 0$. (c) $\Delta u_\theta < 0$, $N_\theta < 0$.

Figure 7.8

Figure 7.8 shows the three elementary dislocations. Their respective stiffness relations are

$$\mu = \frac{EB}{2\pi}\Delta\omega, \quad N_r = \frac{ES}{\pi}\Delta u_r, \quad N_\theta = \frac{ES}{\pi}\Delta u_\theta. \tag{7.95}$$

C. Bending of beams with constant curvature. The results just derived pertain directly to the bending of a beam of rectangular section $t\times(b-a)$, of constant curvature, subtending a full circular arc. But they also apply to the problem of the strain in curved beams with opening angles less than $2\pi$; any two radial sections may be regarded as terminal. The solution will be exact for those cases where the tractions or displacements at the terminal sections are specified as anticipated by the solution. We thus have a theory, analogous to Saint-Venant's, for extension and bending with shear forces but no transverse distributed loads. The main new feature is the occurrence of radial stress between fibers, necessary for their equilibrium.

With an eye toward establishing a one-dimensional formulation, we shall now derive relations between the stresses and the resultant force and moment in any cross section. Airy's function is given by (7.90) in terms of the functions $P(\rho)$ and $F(\rho)$. With $Q(\rho)$ defined by

$$Q(\rho) = \frac{1}{\rho}(\rho P')' = P'' + \frac{1}{\rho}P' = \frac{4\beta^2 \ln\beta}{\beta^2-1} - 2 - 4\ln\rho \tag{7.96}$$

one has

$$(\rho Q')' = 0. \tag{7.97}$$

By construction, also,

$$P(1) = 0, \quad P'(1) = 0, \quad P'(\beta) = 0. \tag{7.98}$$

Similarly, with $G(\rho)$ defined by

$$G(\rho) = \frac{1}{\rho}(\rho F')' - \frac{F}{\rho^2} = F'' + \left(\frac{F}{\rho}\right)' = \frac{4}{\rho} - \frac{8}{\beta^2+1}\rho, \tag{7.99}$$

one has

$$(\rho G')' - \frac{1}{\rho}G = 0, \tag{7.100}$$

and because $(F/\rho)'$ vanishes for $\rho = 1$ and $\rho = \beta$,

$$F(1) = 0, \quad F'(1) = 0, \quad F(\beta) = \beta F'(\beta). \tag{7.101}$$

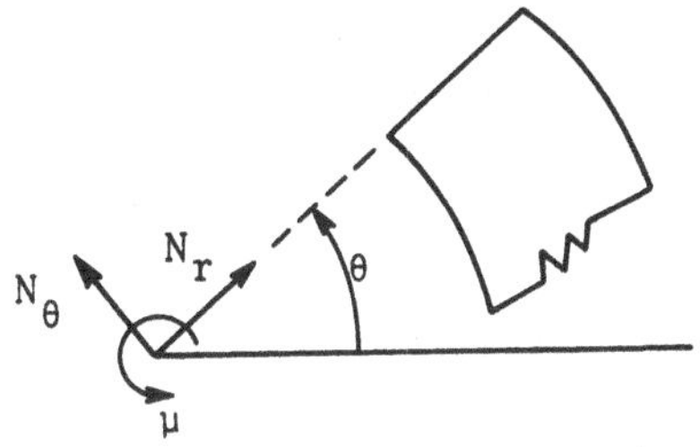


Figure 7.9

From

$$\tau_{r\theta} = -\frac{\gamma\sin\theta + \delta\cos\theta}{a}\left(\frac{F}{\rho}\right)'$$

we calculate the shear force

$$N_r = at\int_1^\beta \tau_{r\theta}\,d\rho = -t(\gamma\sin\theta + \delta\cos\theta)\frac{F(\beta)}{\beta}. \tag{7.102}$$

From

$$\sigma_\theta = -\eta P'' - \frac{\gamma\cos\theta - \delta\sin\theta}{a}F''$$

we calculate the normal force

$$N_\theta = at\int_1^\beta \sigma_\theta\,d\rho = -t(\gamma\cos\theta - \delta\sin\theta)F'(\beta) \tag{7.103}$$

and the moment about the origin

$$\mu = a^2t\int_1^\beta \sigma_\theta\rho d\rho = -a^2t\eta(\rho P'-P)\Big|_1^\beta - at(\gamma\cos\theta-\delta\sin\theta)(\rho F'-F)\Big|_1^\beta$$
$$= a^2t\eta P(\beta). \tag{7.104}$$

From our earlier results, we find

$$F'(\beta) = 2\ln\beta - 2\frac{\beta^2-1}{\beta^2+1} = \frac{8}{t}S$$

$$P(\beta) = \frac{\beta^2-1}{2} - \frac{2\beta^2}{\beta^2-1}(\ln\beta)^2 = \frac{4}{a^2t}B.$$

The stresses in terms of the resultants are

$$\tau_{r\theta} = \frac{N_r}{8aS}\left(\frac{F}{\rho}\right)', \qquad \sigma_\theta = -\frac{\mu}{4B}P'' + \frac{N_\theta}{8aS}F'',$$
$$\sigma_r = -\frac{\mu}{4B}\frac{P'}{\rho} + \frac{N_\theta}{8aS}\left(\frac{F}{\rho}\right)'. \qquad (7.105)$$

Now we calculate the strain energy of the beam per unit angle. For an isotropic material, it is

$$L_\theta = \frac{t}{2}\int_a^b (\sigma_\theta\varepsilon_\theta + \sigma_r\varepsilon_r + \tau_{r\theta}\gamma_{r\theta})rdr$$
$$= \frac{a^2t}{2E}\int_1^\beta (\sigma_\theta+\sigma_r)^2\rho d\rho + \frac{(1+\nu)a^2t}{E}\int_1^\beta (\tau_{r\theta}^2-\sigma_r\sigma_\theta)\rho d\rho.$$

In the first term of the latter form, the harmonic solution

$$\sigma_\theta + \sigma_r = -\frac{\mu}{4B}Q(\rho) + \frac{N_\theta}{8aS}G(\rho)$$

introduces the integrals

$$\int_1^\beta \rho Q^2 d\rho = \int_1^\beta (\rho QP''+QP')d\rho = \rho QP' + QP\Big|_1^\beta - \int_1^\beta [\rho Q'P'+(PQ)']d\rho$$
$$= \rho QP' - \rho Q'P\Big|_1^\beta + \int_1^\beta P(\rho Q')'\, d\rho$$
$$= -\beta Q'(\beta)P(\beta) = 4P(\beta),$$
$$\int_1^\beta \rho G^2 d\rho = \int_1^\beta \rho G[F''+(\frac{F}{\rho})']d\rho = \rho GF'+GF\Big|_1^\beta - \int_1^\beta (F'+\frac{F}{\rho})(\rho G)'d\rho$$
$$= \rho GF' + GF - F(\rho G)'\Big|_1^\beta + \int_1^\beta F[(\rho G')' - \frac{G}{\rho}]d\rho$$
$$= \beta[G(\beta)F'(\beta) - F(\beta)G'(\beta)] = 8F'(\beta),$$

where we have used (7.97) and (7.98) for the first calculation, and (7.100) and (7.101) for the second. Also,

$$\int_1^\beta \rho GQ d\rho = 16H$$

where

$$H = \frac{11}{18}(\beta-1) + \frac{1}{9}\frac{\beta(\beta-1)}{\beta^2+1} - \frac{\beta\ell n\beta}{\beta+1} - \frac{2\beta^2\ell n\,\beta}{3(\beta^2+1)(\beta+1)},$$

an expression which does not seem reducible to a form as simple as the others. In the second integral in the latter form of $L_\theta$, the term in $\tau_{r\theta}^2$ brings in

$$J \equiv \int_1^\beta \rho(\frac{F}{\rho})'(\frac{F}{\rho})'d\rho = 4\ell n\beta\ [1 + \frac{2\beta^2}{(\beta^2+1)^2}] - 6\ \frac{\beta^2-1}{\beta^2+1}\ .$$

In the product $\sigma_r\sigma_\theta$ the term in $N_\theta^2$ involves an integral also having the value $J$, because

$$\int_1^\beta (\frac{F}{\rho})'[\rho(\frac{F}{\rho})'-\rho F'']d\rho = -\int_1^\beta \rho(\frac{F}{\rho})'[\rho(\frac{F}{\rho})']'d\rho = -\frac{1}{2}[\rho(\frac{F}{\rho})']^2\Big|_1^\beta = 0.$$

The term in $\mu^2$ of the same product gives no contribution, because

$$\int_1^\beta P'P''d\rho = \frac{1}{2}(P')^2\Big|_1^\beta = 0.$$

What remains for the second integral is the contribution of the product $\mu N_\theta$:

$$\int_1^\beta [F''P'+\rho P''(\frac{F}{\rho})']d\rho = \int_1^\beta [(P'F')' - \frac{FP''}{\rho}]d\rho$$

$$= P'(F' - \frac{F}{\rho})\Big|_1^\beta + \int_1^\beta P'(\frac{F}{\rho})'d \ = 32K$$

where

$$K = \frac{\beta^3-1}{9(\beta^2+1)} - \frac{\beta\ell n\ \beta}{3(\beta+1)}\ .$$

Thus the strain energy per unit angle becomes

$$L_\theta = \frac{\mu^2}{2EB} - \frac{at\mu N_\theta}{2EBS}[H-2(1+\nu)K] + \frac{N_\theta^2}{2ES} + \frac{(1+\nu)tJ}{64ES^2}(N_r^2 - N_\theta^2).$$

This structure of the energy suggests the establishment of a one-dimensional model of a curved beam based on the concept of a neutral fiber. If the fiber is at $r = R$, there will be a relation $\mu = M + RN_\theta$ between the bending moment $M$, calculated at the level of the neutral fiber, and $\mu$, calculated at the level of the center of curvature. The radius of the neutral fiber may thus be determined by requiring that the energy be a simple sum of energies due respectively to the bending moment, the shear force, and the normal force. Disappearance of the term in $MN_\theta$ requires that

$$\frac{R}{a} = \frac{t}{2S}\hat{H} = \frac{4\hat{H}}{F'(\beta)}, \qquad \hat{H} = H - 2(1+\nu)K, \tag{7.106}$$

thereby giving the energy per unit arc of the neutral fiber the form

$$L = \frac{1}{R} L_\theta = \frac{M^2}{2EI} + \frac{N_\theta^2}{2ES_\theta} + \frac{N_r^2}{2GS_r}, \tag{7.107}$$

where

$$I = BR = \frac{a^3 t}{4}\frac{R}{a} P(\beta), \tag{7.108}$$

$$\frac{at}{S_r} = \frac{J}{4\hat{H}F'(\beta)}, \tag{7.109}$$

$$\frac{at}{S_\theta} = 2\left(\frac{1}{\hat{H}} - \frac{2}{P(\beta)}\frac{R}{a}\right) - 2(1+\nu)\frac{at}{S_r}. \tag{7.110}$$

The numerical values displayed in the following table show that the neutral fiber has a radius slightly less than the arithmetic mean of the extreme radii, that the section resisting the normal stresses is practically equal to the real transversal section, and that the section resisting shear amounts to about 5/6 of it. The calculations were made with $\nu = 0.3$.

| $\beta$ | $R/a$ | $at/S_\theta$ | $at/S_r$ | $a^3t/I$ | $12/(\beta-1)^3$ |
|---|---|---|---|---|---|
| 1.1 | 1.04892 | 10.0145 | 12.0056 | 12 016.0 | 12 000 |
| 1.3 | 1.14090 | 3.35783 | 3.99777 | 449.123 | 444.444 |
| 1.5 | 1.22719 | 2.01065 | 2.40766 | 98.3154 | 96.000 |

As shown by the comparison values in the last column, the moment of inertia is very nearly that of a straight beam with the same section.

We must still ascertain which displacements and generalized strains are associated with the resultants. For this purpose we return to Clapeyron's interior theorem with the strains expressed by formulas (7.69):

$$2L_\theta = t\int_a^b [\sigma_\theta(\frac{u_r}{r} + \frac{1}{r}\frac{\partial u_\theta}{\partial\theta}) + \tau_{r\theta}(-\frac{u_\theta}{r} + \frac{\partial u_\theta}{\partial r} + \frac{1}{r}\frac{\partial u_r}{\partial\theta}) + \sigma_r \frac{\partial u_r}{\partial r}]rdr.$$

Since the boundaries $r = a$ and $r = b$ are not loaded, this expression can be reduced to

$$2L_\theta = t\frac{d}{d\theta}\int_a^b (\sigma_\theta u_\theta + \tau_{r\theta}u_\theta)dr,$$

which has a direct physical interpretation and results from the preceding by integrations by parts. Substitution from (7.105) thus leads to

$$2L_\theta = at\frac{d}{d\theta}\int_1^\beta [-\frac{\mu}{4B}P''u_\theta + \frac{N_\theta}{8aS}F''u_\theta + \frac{N_r}{8aS}(\frac{F}{\rho})'u_r]d\rho. \qquad (7.111)$$

This expression will now be compared with one furnished by a one-dimensional theory for a fiber of radius $R$, with radial and tangential displacements denoted respectively by $U_r$ and $U_\theta$. The virtual work of the forces applied to the extremities of a segment $R\,d\theta$ of fiber will be

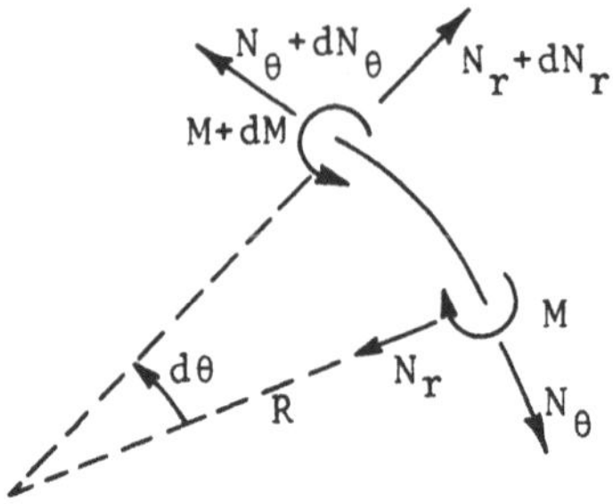


Figure 7.10

$$\frac{d}{d\theta}\,(M\delta\alpha + N_\theta \delta U_\theta + N_r \delta U_r)\;d\theta \tag{7.112}$$

where $\alpha$ is the rotation of the cross section. The expression to compare with (7.111) is therefore

$$\frac{d}{d\theta}\,(M\alpha + N_\theta U_\theta + N_r U_r)$$

with $\mu = M + RN_\theta$. The comparison furnishes definitions of the displacements of the neutral fiber and of the sectional rotation in terms of weighted means of the exact displacements:

$$\alpha = -\frac{1}{P(\beta)}\int_1^\beta P''u_\theta d\rho = \frac{\int_a^b [1+\ln\frac{r}{a} - \frac{b^2}{b^2-a^2}(1+\frac{a^2}{r^2})\ln\frac{b}{a}]u_\theta dr}{\frac{b^2-a^2}{4} - \frac{a^2b^2}{b^2-a^2}(\ln\frac{b}{a})^2},$$

$$U_r = \frac{1}{F'(\beta)}\int_1^\beta (\frac{F}{\rho})'u_r d\rho = \frac{\int_a^b \frac{(r^2-a^2)(r^2-b^2)}{(b^2+a^2)r^3}\,u_r\,dr}{\frac{b^2-a^2}{b^2+a^2} - \ln\frac{b}{a}},$$

$$U_\theta - R\alpha = \frac{1}{F'(\beta)}\int_1^\beta F''u_\theta d\rho = \frac{\int_a^b [\frac{3r}{a^2+b^2} - \frac{1}{r} - \frac{a^2b^2}{(b^2+a^2)r^3}]u_\theta dr}{\frac{b^2-a^2}{b^2+a^2} - \ln\frac{b}{a}}.$$

One may verify that these definitions reduce to identities when we take respectively $u_\theta = r\alpha$, $u_r = U_r$, $u_\theta = U_\theta$.

We now develop the consequences of (7.112). The virtual work should vanish for the translations

$$\delta U_r = \delta U\cos\theta + \delta V\sin\theta, \quad \delta U_\theta = -\delta U\sin\theta + \delta V\cos\theta, \quad \delta\alpha = 0$$

with $\delta U$ and $\delta V$ constant. This confirms the equilibrium equations in terms of the resultants

$$dN_r/d\theta - N_\theta = 0, \quad dN_\theta/d\theta + N_r = 0, \tag{7.113}$$

which may be found from equations (7.102) and (7.103).

The virtual work should also vanish for the small rigid rotation

$$\delta U_r = 0, \quad \delta U_\theta = R\delta\Omega, \quad \delta\alpha = \delta\Omega$$

with $\delta\Omega$ constant, and this yields the third equilibrium equation

$$\frac{d}{d\theta}(M + RN_\theta) = \frac{d\mu}{d\theta} = 0, \tag{7.114}$$

confirmed by the earlier result (7.104). When simplified by the equilibrium equation, the virtual work

$$M\delta(\frac{d\alpha}{d\theta}) + N_\theta\delta(U_r + \frac{d}{d\theta}U_\theta) + N_r\delta(-U_\theta + R\alpha + \frac{d}{d\theta}U_r)$$

displays the generalized deformations $d\alpha/d\theta$ conjugate to the bending moment $M$, $U_r + dU_\theta/d\theta$ conjugate to the normal stress $N_\theta$, and $R\alpha - U_\theta + dU_r/d\theta$ conjugate to the shear force $N_r$. Thus Clapeyron's interior theorem for the one-dimensional model may be written

$$2L_\theta = M\frac{d\alpha}{d\theta} + N_\theta(U_r + \frac{d}{d\theta}U_\theta) + N_r(-U_\theta + R\alpha + \frac{d}{d\theta}U_r)$$

and leads, by comparison with (7.107), to generalized constitutive equations for the theory of curved beams

$$\frac{d\alpha}{d\theta} = \frac{RM}{EI}, \qquad U_r + \frac{dU_\theta}{d\theta} = \frac{RN_\theta}{ES_\theta},$$
$$R\alpha - U_\theta + \frac{dU_r}{d\theta} = \frac{RN_r}{GS_r}. \tag{7.115}$$

As with straight beams, one may study the relative importance of deformations caused by shear and normal force. We study a curved beam perfectly clamped at the section $\theta = 0$ and loaded at the section $\theta = \psi$ (Figure 7.11).

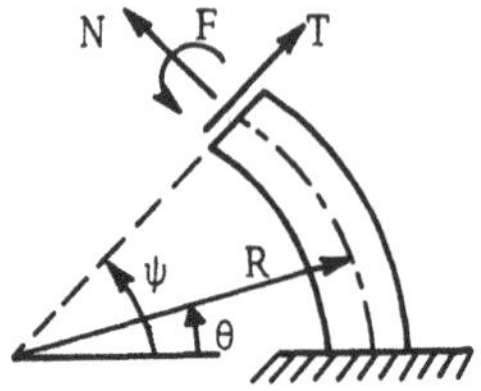


Figure 7.11

Since

$$d^2N_r/d\theta^2 + N_r = 0,$$

the shear force in an arbitrary section is of the form $N_r = A\cos\theta + B\sin\theta$ where $N_\theta = dN_r/d\theta = -A\sin\theta + B\cos\theta$. Adjusting the constants for the terminal loads yields

$$N_r = T\cos(\theta-\psi) + N\sin(\theta-\psi), \quad N_\theta = -T\sin(\theta-\psi) + N\cos(\theta-\psi).$$

Therefore, since $M + RN_\theta = F + RN$, the bending moment is

$$M = F + RN[1 - \cos(\theta-\psi)] + RT\sin(\theta-\psi).$$

Integration of the rotation of the section with the boundary condition $\alpha(0) = 0$ yields

$$EI\alpha = FR\theta + R^2N[\theta - \sin\psi - \sin(\theta-\psi)] + R^2T[\cos\psi - \cos(\theta-\psi)].$$

By taking

$$U_\theta = P\cos\theta + Q\sin\theta, \quad U_r = P\sin\theta - Q\cos\theta$$

we can obtain the displacements by the method of variation of constants. With $P$ and $Q$ constant this is the general solution of the last two of equations (7.115), rendered homogeneous by taking $N_\theta = 0$, $N_r = 0$, and $\alpha = 0$. Considering $P$ and $Q$ now as functions of $\theta$, we satisfy the inhomogeneous equations by integrating the relations

$$\frac{dP}{d\theta} = \frac{RN_\theta}{ES_\theta}\cos\theta + \left(\frac{RN_r}{GS_r} - R\alpha\right)\sin\theta, \quad \frac{dQ}{d\theta} = \frac{RN_\theta}{ES_\theta}\sin\theta - \left(\frac{RN_r}{GS_r} - R\alpha\right)\cos\theta,$$

with the boundary conditions $P(0) = 0$ and $Q(0) = 0$. The result may be written in the form

$$U_\theta - iU_r = \frac{R}{ES_\theta}\int_0^\theta N_\theta(\omega)e^{i(\omega-\theta)}d\omega - \frac{iR}{GS_r}\int_0^\theta N_r(\omega)e^{i(\omega-\theta)}d\omega + iR\int_0^\theta \alpha(\omega)e^{i(\omega-\theta)}d\omega.$$

By substituting the expressions found for $N_\theta$, $N_r$, and $\alpha$, and integrating, we obtain

$$\begin{aligned} U_\theta = {} & \frac{R^2(F+RN)}{EI}(\theta-\sin\theta) + \frac{R^3}{EI}(T\cos\psi - N\sin\psi)(1-\cos\theta) \\ & - \frac{R}{2}\left(\frac{R^2}{EI} + \frac{1}{GS_r} - \frac{1}{ES_\theta}\right)(T\sin\psi + N\cos\psi)\sin\theta \\ & + \frac{R}{2}\left(\frac{R^2}{EI} + \frac{1}{GS_r} + \frac{1}{ES_\theta}\right)[T\theta\sin(\psi-\theta) + N\theta\cos(\psi-\theta)], \\ U_r = {} & \frac{R^2(F+RN)}{EI}(\cos\theta - 1) \\ & - \frac{R}{2}\left(\frac{R^2}{EI} - \frac{1}{GS_r} + \frac{1}{ES_\theta}\right)(T\cos\psi - N\sin\psi)\sin\theta \\ & + \frac{R}{2}\left(\frac{R^2}{EI} + \frac{1}{GS_r} + \frac{1}{ES_\theta}\right)[T\theta\cos(\psi-\theta) - N\theta\sin(\psi-\theta)]. \end{aligned}$$

These expressions suggest that if

$$\frac{1}{GS_r} + \frac{1}{ES_\theta} << \frac{R^2}{EI}$$

or, in practice,

$$\left(\frac{b-a}{R}\right)^2 << 3,$$

we may neglect the deformations caused by the shear and normal force, thereby simplifying the constitutive equations to

$$\frac{d\alpha}{d\theta} = \frac{RM}{EI}, \quad U_r + \frac{dU_\theta}{d\theta} = 0, \quad R\alpha - U_\theta + \frac{dU_r}{d\theta} = 0.$$

This leads to the differential equation

$$\frac{d^2U_r}{d\theta^2} + U_r = -\frac{R^2M}{EI}.$$

A curved beam meeting the above restriction will be called "thin". For such a beam the displacements and the rotation of the extremity $\theta = \psi$ reduce to

$$\alpha = F\frac{R}{EI}\psi + N\frac{R^2}{EI}(\psi - \sin\psi) + T\frac{R^2}{EI}(\cos\psi - 1),$$

$$U_\theta = F\frac{R^2}{EI}(\psi - \sin\psi) + N\frac{R^3}{EI}\left(\frac{3\psi}{2} - 2\sin\psi + \frac{\sin 2\psi}{4}\right) + T\frac{R^3}{EI}\left(-\frac{3}{4} + \cos\psi - \frac{\cos 2\psi}{4}\right),$$

$$U_r = F\frac{R^2}{EI}(\cos\psi - 1) + N\frac{R^3}{EI}\left(-\frac{3}{4} + \cos\psi - \frac{\cos 2\psi}{4}\right) + T\frac{R^3}{EI}\left(\frac{\psi}{2} - \frac{\sin 2\psi}{4}\right).$$

The corresponding influence functions are shown in the Figure 7.12 as functions of the angle $\psi$. Except for the complete, split ring, for which $\psi = 2\pi$, they never vanish if $\psi > 0$. The same situation would prevail for influence functions which take account of deformations caused by normal and shear forces, and this fact helps to justify the use of the simplified forms.

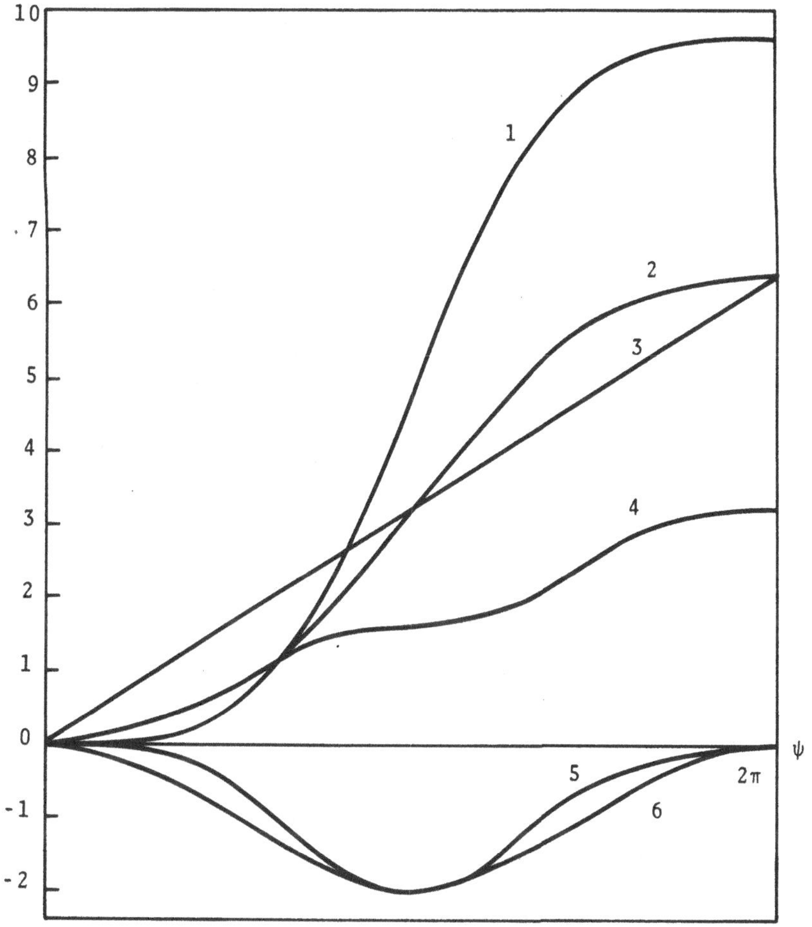


1) $\frac{EIU_\theta}{R^3N} = \frac{3}{2}\psi - \sin\psi + \frac{\sin 2\psi}{4}$

2) $\frac{\alpha EI}{R^2N} = \frac{EIU_\theta}{R^2F} = \psi - \sin\psi$

3) $\frac{EI\alpha}{RF} = \psi$

4) $\frac{EIU_r}{R^3T} = \frac{\psi - \sin\psi\cos\psi}{2}$

5) $\frac{EIU_\theta}{R^3T} = \frac{EIU_r}{R^3N} = -\frac{3}{4} + \cos\psi - \frac{\cos 2\psi}{4}$

6) $\frac{EIU_r}{R^2F} = \frac{EI\alpha}{R^2T} = \cos\psi - 1$

Figure 7.12. Influence functions for the thin circular beam.

The one-dimensional theory admits only simplified boundary conditions, such as the approximation to perfect clamping which we used in the preceding problem; such conditions serve to extend the range of application of the exact solution. The theory also admits an approximate analysis of the deformation when the loads are of a more general type. Although the general formulas of Section 7.10 permit the development of exact solutions for curved beams or rings loaded along the boundaries $r = a$ and $r = b$, the mean displacements $\alpha$, $U_r$, and $U_\theta$, and in certain cases the distribution of the stresses $\sigma_\theta$, can be found with adequate accuracy by using only equivalent loads distributed along the neutral fiber. The radial and tangential distributed loads and the distributed couple may vary with $\theta$, and are given respectively by

$$p_r = t\,\frac{b}{R}\,\sigma_r(b) - t\,\frac{a}{R}\,\sigma_r(a),$$

$$p_\theta = t\,\frac{b}{R}\,\tau_{r\theta}(b) - t\,\frac{a}{R}\,\tau_{r\theta}(a),$$

$$c + Rp_\theta = t\,\frac{b^2}{R}\,\tau_{r\theta}(b) - t\,\frac{a^2}{R}\,\tau_{r\theta}(a).$$

The static equivalence of these with the real loads results from equating the virtual work of the latter, acting through rigid virtual displacements of a segment $d\theta$ of the beam, with that of the distributed loads, acting through rigid virtual displacements of the neutral fiber.

To incorporate the effect of the distributed loads, one need only add to the virtual work (7.112) the expression

$$(p_r\delta U_r + p_\theta\delta U_\theta + c\delta\alpha)\,R\delta\theta.$$

There follow the generalized equilibrium equations

$$dN_r/d\theta - N_\theta + p_r R = 0, \quad dN_\theta/d\theta + N_r + p_\theta = 0, \qquad (7.113)'$$

$$\frac{d}{d\theta}(M + RN_\theta) + c + Rp_\theta = 0. \qquad (7.114)'$$

As with the straight beams, the approximate solutions based on these more general equations of equilibrium and on the constitutive equations (7.115) admit variational justification.

When the beam is thin, all loadings leading to the same distribution of the bending moment become equivalent, and this circumstance may be helpful in certain mechanical problems. A segment of a piston, for example, may be regarded as a split rectangular ring and should take a circular form with diameter equal to that of the cylinder when it is subjected to a uniform radial pressure $\sigma_r(b)$ (Figure 7.13).

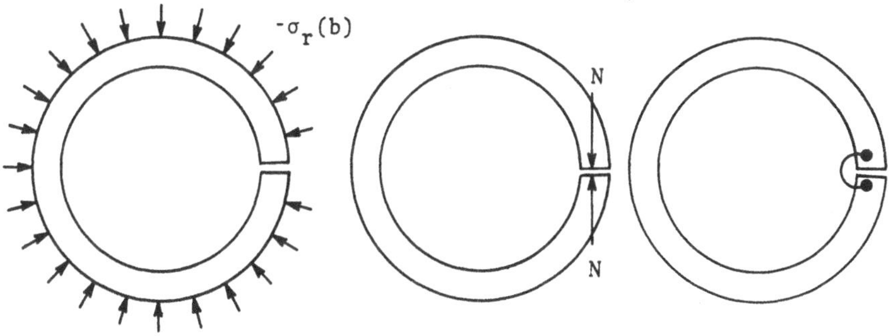


Figure 7.13

Let $p = -p_r$ be the radial load of the neutral fiber. The bending moment then has the distribution

$$M = pR^2(1 - \cos\theta),$$

which is equivalent to that generated by a pair of normal forces $N = pR$ applied to the opposing faces. In practice, the two ends are held by an articulated joint exerting the reactions N in question. The separation between the terminal sections in the unstressed configuration is connected with the applied pressure by the relation

$$\Delta U_\theta = 3\pi \frac{R^4}{EI} p.$$

The dynamometric ring is another example of an application of the one-dimensional theory of curved beams (Figure 7.14).

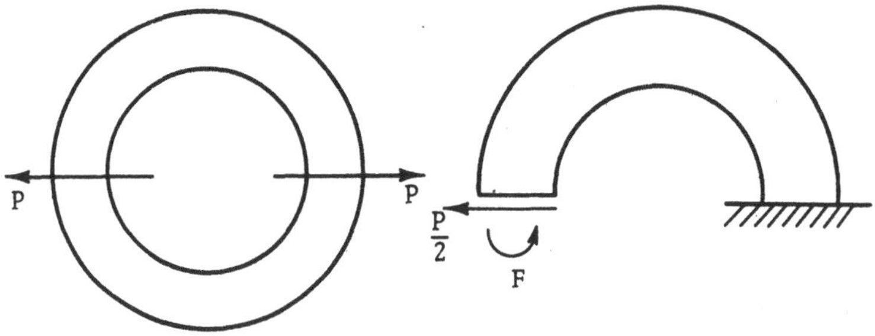


Figure 7.14

The ring is complete, and is subjected to two concentrated and diametrically opposed radial forces P. By symmetry, this is equivalent to the problem of a half-ring with one end perfectly clamped and the other subjected to a shear force P/2 and a bending moment F. The latter is to be adjusted so that the rotation of the terminal section vanishes.

With the notation of the preceding problem we may write $\psi = \pi$, $N = 0$, $T = P/2$. Since

$$EI\alpha = RF\theta + R^2 P \frac{1}{2} (\cos\theta - 1),$$

the vanishing of the rotation for $\theta = \pi$ yields

$$F = PR/\pi = \mu.$$

The distribution of the bending moment then becomes

$$M = F - RT \sin(\pi - \theta) = PR\left(\frac{1}{\pi} - \frac{\sin\theta}{2}\right),$$

while the shear and normal forces have the expressions

$$N_r = -\frac{P}{2}\cos\theta, \quad N_\theta = \frac{P}{2}\sin\theta.$$

The stresses are completely determined by these with (7.91), (7.92), and (7.105).

The radial displacement of the end is

$$U_r = \frac{PR^3}{EI}\left(\frac{\pi}{4} - \frac{2}{\pi}\right) + \frac{\pi PR}{4}\left(\frac{1}{ES_\theta} + \frac{1}{GS_r}\right),$$

which is the stiffness relation for the ring.

If we measure the transverse contraction of the ring instead of its extension, the theory predicts the value

$$\frac{PR^3}{EI}\left(\frac{2}{\pi} - \frac{1}{2}\right) + \frac{PR}{2}\left(\frac{1}{GS_r} - \frac{1}{ES_\theta}\right).$$

Because $\pi/4 - 2/\pi = 0.14878$ while $2/\pi - 1/2 = 0.13662$, the latter measure is slightly less sensitive than the former.

D. The annular ring loaded by shear tractions.

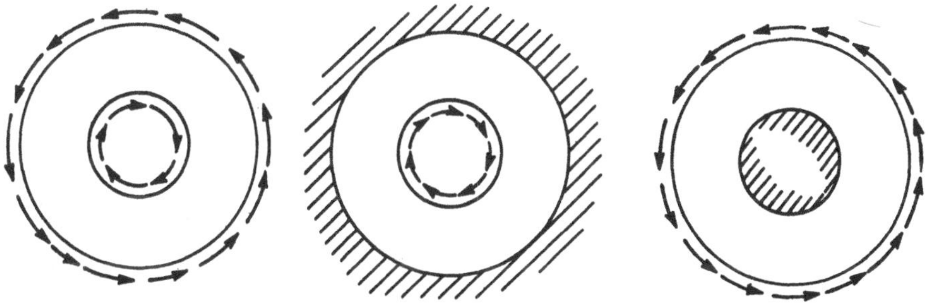

Figure 7.15

This particular case of the ring loaded along its borders $r = a$ and $r = b$ is remarkable because while the state of stress depends only on $r$, Airy's function varies with $\theta$ and may even be multivalent, because the resultant of the stresses in the cavity is not zero but a couple with intensity which will be denoted by $C$. The solution is supplied by the first term of (7.79), $\phi = \beta\theta$, which yields $\sigma_r = \sigma_\theta = 0$, $\tau_{r\theta} = \beta r^{-2}$, $u_r = 0$, $2Gu_\theta = -\beta r^{-1}$. It follows that

$$C = t\int_0^{2\pi} \tau_{r\theta} r^2 d\theta = 2\pi\beta t.$$

Since the radial displacement vanishes identically, one may always add a rotation of the whole in order to annul the tangential displacement, either at $r = b$, corresponding to the ring loaded in torsion on the inside and perfectly clamped on the outside, or at $r = a$, if it is loaded on the outside and clamped on the inside.

The relative rotation between the fibers $r = b$ and $r = a$ is

$$\frac{u_\theta(b)}{b} - \frac{u_\theta(a)}{a} = \frac{C}{G}\frac{b^2 - a^2}{4\pi t a^2 b^2}$$

which is the stiffness relation for this type of loading.

If on the other hand we apply the generalized one-dimensional theory, we find for the equivalent loads on neutral fiber

$$p_r = 0, \quad p_\theta R = -t\beta\,\frac{b-a}{ab}, \quad c + p_\theta R = 0,$$

and for solutions of the equilibrium equations (7.113)' and (7.114)' and the constitutive equations (7.115)

$$N_\theta = 0, \quad M = 0, \quad N_r = t\beta\,\frac{b-a}{ab},$$

$$U_r = 0, \quad U_\theta = -\frac{R}{GS_r}N_r + R\alpha_0, \quad \alpha = \alpha_0 .$$

The results $\sigma_\theta = 0$ and $\sigma_r = 0$ which follow for the stress field are correct, but the distribution of the shear stresses is completely different. This is not at all surprising, because the boundary conditions $\tau_{r\theta} = 0$ for $r = a$ and $r = b$, intrinsic to the one-dimensional theory, are essentially different from the real conditions.

E. The thick tube under pressure.

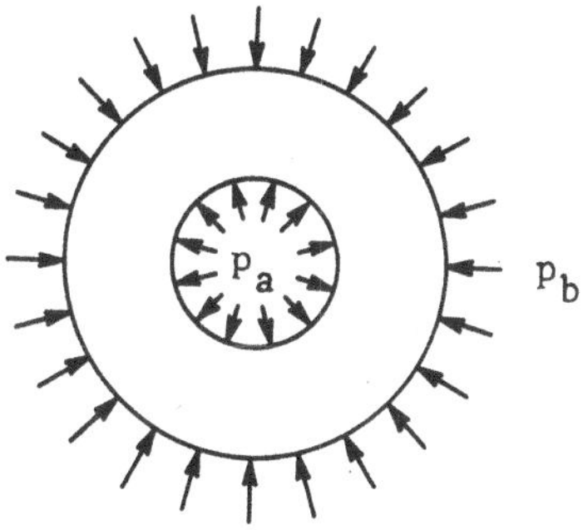


Figure 7.16

We begin study of this problem with the case of plane strain, i.e., $\varepsilon_z = 0$. Although the cavity is loaded, the internal resultant vanishes and Airy's function is univalent; indeed, it depends only on $r$ and thus is of the form (7.87). Formulas (7.71) yield

$$\tau_{r\theta} = 0, \qquad \sigma_r = Br^{-2} + 2C + D(1+2\ell n\ r),$$
$$\sigma_\theta = -Br^{-2} + 2C + D(3+2\ell n\ r).$$

With reference to the general solution (7.79) for Airy's function in a ring, we have the correspondence

$$A = -\alpha_o, \quad B = -\alpha, \quad C = \lambda_1 - \lambda, \quad D = \lambda.$$

As we have seen, the tangential displacement associated with $\lambda$ is multivalent. We therefore conclude that $\lambda = D = 0$.

This result can alternatively be obtained by observing that the central symmetry of the problem demands $u_\theta \equiv 0$. Then $\varepsilon_r = du_r/dr$, $\varepsilon_\theta = u_r/r$, and therefore

$$\varepsilon_r = \frac{d}{dr}\,(r\varepsilon_\theta).$$

Now, by (7.17), $\hat{E}\varepsilon_r = \sigma_r - \hat{\nu}\sigma_\theta$ and $\hat{E}\varepsilon_\theta = \sigma_\theta - \hat{\nu}\sigma_r$, which allow the preceding relation to be expressed in the form

$$\sigma_r - \hat{\nu}\sigma_\theta = \frac{d}{dr}(r\sigma_\theta - \hat{\nu}r\sigma_r).$$

When we substitute the preceding solution into this equation, the terms in B and C vanish, and we may satisfy it only by taking $D = 0$.

The boundary conditions

$$-p_a = \frac{B}{a^2} + 2C, \qquad -p_b = \frac{B}{b^2} + 2C$$

supply the values of the constants B and C. For the stresses we find

$$\sigma_r = \frac{a^2p_a - b^2p_b}{b^2 - a^2} - \frac{a^2b^2(p_a-p_b)}{b^2 - a^2}\frac{1}{r^2}$$

$$\sigma_\theta = \frac{a^2p_a - b^2p_b}{b^2 - a^2} - \frac{a^2b^2(p_a-p_b)}{b^2 - a^2}\frac{1}{r^2}.$$

For the radial displacement, we have directly

$$\hat{E}u_r = r\hat{E}\varepsilon_\theta = r(\sigma_\theta - \hat{\nu}\sigma_r)$$

or, after substituting the stresses and using (7.18),

$$2Gu_r = (1-2\nu)\frac{a^2p_a - b^2p_b}{b^2 - a^2}r + \frac{a^2b^2(p_a-p_b)}{b^2 - a^2}\frac{1}{r}.$$

Along the axis of the tube, the stress $\sigma_z$ takes the constant value

$$\sigma_z = \nu(\sigma_r + \sigma_\theta) = 2\nu\frac{a^2p_a - b^2p_b}{b^2 - a^2}.$$

This circumstance allows discussing the same problem for the case of plane stress, when the tube is free to elongate or contract and $\sigma_z \equiv 0$. The field of radial and tangential stresses is unchanged, but the constitutive equations which now apply,

$$E\varepsilon_r = \sigma_r - \nu\sigma_\theta, \quad E\varepsilon_\theta = \sigma_\theta - \nu\sigma_r ,$$

lead to an adjusted radial displacement

$$Eu_r = r(\sigma_\theta - \nu\sigma_r) = (1-\nu)\frac{a^2p_a - b^2p_b}{b^2-a^2}r + (1+\nu)\frac{a^2b^2(p_a-p_b)}{b^2-a^2}\frac{1}{r} .$$

The tube undergoes a uniform axial elongation

$$E\varepsilon_z = -\nu(\sigma_r+\sigma_\theta) = -2\nu\frac{a^2p_a - b^2p_b}{b^2 - a^2} .$$

F. <u>Concentric cylindrical tubes and rings</u>. We assume

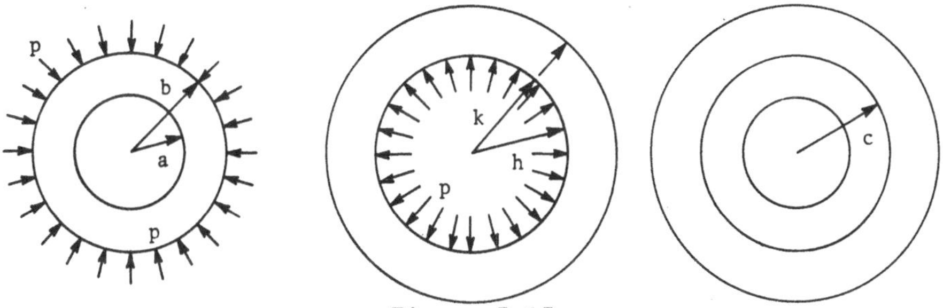


Figure 7.17

the tubes free to contract or elongate. According to the results of the preceding section, when subjected to an exterior pressure $p$ the interior tube undergoes an exterior radial displacement

$$E_i(u_r)_i = (\nu_i - \frac{b^2 + a^2}{b^2 - a^2})pb.$$

Subjected to internal pressure $p$, the exterior tube undergoes an interior radial displacement

$$E_e(u_r)_e = (\nu_e + \frac{k^2 + h^2}{k^2 - h^2})ph.$$

Let the pressure $p$ be such that the interior tube fits snugly into the exterior tube. Then

$$b + (u_r)_i = h + (u_r)_e = c.$$

The first equality supplies the relation between the interface pressure and the radial interference b-h, or the difference in radii in the unloaded state:

$$b-h = p\{\frac{b}{E_i}(-\nu_i + \frac{b^2+a^2}{b^2-a^2}) + \frac{h}{E_e}(\nu_e + \frac{k^2+h^2}{k^2+h^2})\}.$$

When the materials are the same, this simplifies to

$$b-h = \frac{p}{E}\{\nu(h-b) + h\,\frac{k^2+h^2}{k^2-h^2} + b\,\frac{b^2+a^2}{b^2-a^2}\}.$$

Since $h-b$ is negligible compared with $b$, the final result is the simple formula

$$b-h = \frac{2p}{E}\,b^3\,\frac{k^2-a^2}{(k^2-h^2)(b^2-a^2)}\,.$$

This problem illustrates another possibility of the existence of internal stresses which cannot be relaxed by a change of configuration. They may exist even when the region is simply connected ($a = 0$). They differ fundamentally from those due to the closing of Volterra's dislocations, because they violate the conditions of regularity: on the interface the discontinuity of $\sigma_\theta$ causes a discontinuity in the values of $\varepsilon_r$ and $\varepsilon_\theta$.

G. Force concentrated at the origin in an infinite plate. From the general solution presented in Section 7.10, we keep the terms which generate stresses inversely proportional to $r$. They are

$$H = (\gamma+i\delta)\zeta(\ell n\zeta-1), \quad (1+\nu)F = (\lambda+i\mu)\ell n\,\zeta.$$

In order that the displacement field remain univalent, it is necessary to choose

$$\gamma + i\delta = \frac{3-\nu}{1+\nu} (\lambda - i\mu)$$

and up to a factor, as we shall see, the coefficients $\lambda$ and $\mu$ are the components of the vectorial sum of the forces $P_x$ and $P_y$ acting, per unit thickness, on a cavity containing the origin.

By using (7.41) we obtain the univalent displacement field

$$2G(u+iv) = 2 \frac{3-\nu}{1+\nu} (\lambda+i\mu)\ell n\ r - (\lambda-i\mu)e^{2i\theta}$$

while the formulas (7.77) in the case of generalized plane stress yield

$$\sigma_r + \sigma_\theta + iE\omega = 4(\lambda + i\mu) \frac{e^{-i\theta}}{r}$$

$$\sigma_r-\sigma_\theta-2i\tau_{r\theta} = 2(\lambda+i\mu) \frac{e^{-i\theta}}{r} + 2 \frac{3-\nu}{1+\nu} (\lambda-i\mu) \frac{e^{i\theta}}{r}$$

or

$$\sigma_r = \frac{2}{1+\nu}(3+\nu) \frac{\lambda\cos\theta + \mu\sin\theta}{r}, \qquad \tau_{r\theta} = \frac{2}{1+\nu}(1-\nu) \frac{\mu\cos\theta - \lambda\sin\theta}{r},$$

$$\sigma_\theta = -\frac{2}{1+\nu}(1-\nu) \frac{\lambda\cos\theta + \mu\sin\theta}{r} .$$

For any circular cavity about the origin we have

$$-P_x = \int_0^{2\pi} (\sigma_r\cos\theta-\tau_{r\theta}\sin\theta)rd\theta, \qquad -P_y = \int_0^{2\pi} (\sigma_r\sin\theta+\tau_{r\theta}\cos\theta)rd\theta$$

or, after substituting the known values and performing the integrations,

$$P_x = -\frac{8\pi}{1+\nu} \lambda, \qquad P_y = -\frac{8\pi}{1+\nu} \mu.$$

The moment with respect to the origin is

$$M_o = \int_0^{2\pi} \tau_{r\theta}r^2\ d\theta = 0,$$

and we may consider the field as that of a concentrated force situated at the origin.

Another way to reach this result is to study the corresponding Airy's function

$$\phi = \mathrm{Re}\{(1+\nu)\bar{\zeta}F-\bar{H}\} = \mathrm{Re}[(\lambda+i\mu)\bar{\zeta}\{\ln\zeta - \frac{3-\nu}{1+\nu}(\ln\bar{\zeta}-1)\}]$$

$$= \frac{\lambda+i\mu}{2}[\bar{\zeta}\ln\zeta - \frac{3-\nu}{1+\nu}\bar{\zeta}(\ln\bar{\zeta}-1)] + \frac{\lambda-i\mu}{2}[\zeta\ln\bar{\zeta} - \frac{3-\nu}{1+\nu}\zeta(\ln\zeta-1)].$$

When one traverses a closed circuit surrounding the origin starting from a point $P$ with label $\zeta$, the increment in $\phi$ is

$$\Delta\phi = \mathrm{Re}[(\lambda+i\mu)\bar{\zeta}2\pi i(1 + \frac{3-\nu}{1+\nu})] = \frac{8\pi}{1+\nu}(\lambda y-\mu x).$$

At the same time, by using the operator $\partial$ of Section 7.1,

$$\partial\phi = \frac{\lambda+i\mu}{2}[2\frac{\bar{\zeta}}{\zeta}] + \frac{\lambda-i\mu}{2}[2\ln\bar{\zeta} - 2\frac{3-\nu}{1+\nu}\ln\zeta].$$

If we traverse the closed circuit this gives

$$\Delta\partial\phi = -2\pi i(\lambda-i\mu)(1+\frac{3-\nu}{1+\nu})$$

or

$$\Delta\frac{\partial\phi}{\partial x} = -\frac{8\pi}{1+\nu}\mu, \qquad \Delta\frac{\partial\phi}{\partial y} = \frac{8\pi}{1+\nu}\lambda.$$

These results confirm those above when one applies to the closed circuit the interpretation of Airy's function and its partial derivatives set forth in Section 7.6.

As to the problem of the displacement caused by this loading, we observe that replacing $r$ by $r/r_0$ as argument of the logarithm merely introduces a translation, making the contribution of this term zero on the circle of chosen radius $r_0$. By adding the field due to the term

$$(1+\nu)F = (\lambda_2 + i\mu_2)\zeta^2$$

we can annihilate on the same circle the term in $e^{2i\theta}$ by the choice

$$\frac{3-\nu}{1+\nu} r_0^2 (\lambda_r + i\mu_r) = \lambda + i\mu,$$

and by adding a new translation we can annihilate $u$ and $v$ completely at $r = r_0$. The final result is

$$2G(u+iv) = 2\frac{3-\nu}{1+\nu}(\lambda+i\mu)\ln\frac{r}{r_0} + (\lambda+i\mu)(\frac{r^2}{r_0^2}-1)e^{2i\theta} - 2\frac{1+\nu}{3-\nu}(\lambda-i\mu)(\frac{r^2}{r_0^2}-1).$$

By thus adjusting the boundary conditions we have annulled the displacements on a circle with a radius as large as we please, and have introduced a correction to the stress field without changing the resultant of the forces acting in the cavity. For a single force $P_x$,

$$\sigma_r = -\frac{P_x}{4\pi}[\frac{3+\nu}{r} + \frac{(1+\nu)^2}{3-\nu}\frac{r}{r_0^2}]\cos\theta,$$

$$\tau_{r\theta} = \frac{P_x}{4\pi}[\frac{1-\nu}{r} - \frac{(1+\nu)^2}{3-\nu}\frac{r}{r_0^2}]\sin\theta,$$

$$\sigma_\theta = \frac{P_x}{4\pi}[\frac{1-\nu}{r} - \frac{3(1+\nu)^2}{3-\nu}\frac{r}{r_0^2}]\cos\theta.$$

The reactions induced at the locked ring $r = r_0$ decrease in inverse proportion to its radius.

It is impossible to find an expression for displacements corresponding to vanishing displacement at infinity. The presence of the logarithmic term causes the local displacement at any fixed $r$ to become infinite as the radius of vanishing displacement $r_0$ tends to infinity.

# Chapter 8
# Bending of Plates

Let the middle surface of a flat plate of thickness $2h$ contain the Cartesian axes $Ox$ and $Oy$. The bounding surfaces are the two faces at $z = \pm h$, and the cylindrical surface defined by a closed directrix $c$ lying in the middle surface. The thickness $2h$ is assumed to be small compared with the transverse dimensions.

Loads which are symmetric with respect to the middle surface have already been discussed (see Chapter 7). Here we consider the anti-symmetric case. General loadings may be obtained by superposition as long as one stays within the context of the linear theory of elasticity.

The principal objective here, as in the last chapter, is to reduce to two dimensions the formulation of a problem which really contains three. This will be possible if the z-dependence of the solution can be made explicit, either from an exact three-dimensional treatment (as in the case of plane strain), or from simplifying rational hypotheses (as in the case of generalized plane stress). The latter point of view will be adopted, and implemented with the aid of a variational principle. We shall find that various types of variational approximations lead to essentially the same two-dimensional equations, the differences occurring only in the values of certain stiffness coefficients.

## 8.1. Basic Hypotheses

The anti-symmetry of bending, and the example of beam theory, suggest the fundamental approximative assumptions

$$\sigma_x = zS_x(x,y), \quad \sigma_y = zS_y(x,y), \quad \tau_{xy} = zS_{xy}(x,y).$$

Each may be regarded as the first term in a Taylor series, in odd powers of $z$, for the exact field.

It is convenient to introduce at once a more direct physical interpretation for the functions of two variables $(x,y)$ appearing above. Let the distributed bending moments and twisting moment be defined by

$$M_x = \int_{-h}^{h} \sigma_x z\, dz, \quad M_y = \int_{-h}^{h} \sigma_y z\, dz, \quad M_{xy} = \int_{-h}^{h} \tau_{xy} z dz. \tag{8.1}$$

When positive, they act as shown in Figure 8.1.

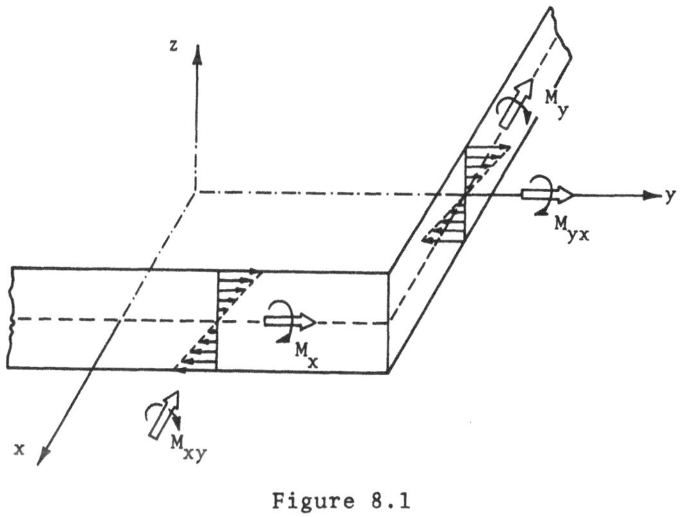


Figure 8.1

Now by using the proposed approximations for the stresses, we find

$$\sigma_x = \frac{3z}{2h^3} M_x(x,y), \quad \sigma_y = \frac{3z}{2h^3} M_y(x,y), \quad \tau_{xy} = \frac{3z}{2h^3} M_{xy}(x,y). \tag{8.2}$$

We further hypothesize that, approximately,

$$\sigma_z \equiv 0, \tag{8.3}$$

and that at $z = \pm h$,

$$\tau_{zx} = \tau_{zy} = 0. \tag{8.4}$$

The faces $z = \pm h$ are thus not externally loaded, and any effective transverse load must be introduced by body forces. Replacement of the latter by the more realistic application of transverse pressures on the faces may be effected by a posteriori corrections.

## 8.2. Application of the Canonical Variational Principle

We shall derive field equations for plate bending by using the general canonical principle

$$\delta(H_1 + H_2 - U + P) = 0,$$

where, in view of (8.3),

$$H_1 = \iiint [\sigma_x \frac{\partial u}{\partial x} + \tau_{xy}(\frac{\partial u}{\partial y} + \frac{\partial v}{\partial x}) + \sigma_y \frac{\partial v}{\partial y}]dxdydz,$$

$$H_2 = \iiint [\tau_{zx}(\frac{\partial w}{\partial x} + \frac{\partial u}{\partial z}) + \tau_{zy}(\frac{\partial w}{\partial y} + \frac{\partial v}{\partial z})]dxdydz,$$

$$U = \frac{1}{2E}\iiint[\sigma_x^2 + \sigma_y^2 - 2\nu\sigma_x\sigma_y + 2(1+\nu)\tau_{xy}^2]dxdydz$$

$$+ \frac{1}{2G}\iiint(\tau_{zx}^2 + \tau_{zy}^2)dxdydz,$$

$$P = -\iiint Zw\, dxdydz - \int_{-h}^{h} dz \oint_{\bar{C}} (\bar{t}_x u + \bar{t}_y v + \bar{t}_z w)ds.$$

The quantity $U$ is the strain energy expressed in terms of stresses, or the complementary strain energy, while $P$ is the potential energy of the applied loads. The integral in the second term of $P$ extends only over that part of the cylindrical boundary where the tractions $\bar{t}_x$, $\bar{t}_y$, $\bar{t}_z$ are specified. The variations in the stresses are arbitrary, while those of the displacements must vanish on that part of the boundary where displacement boundary conditions are imposed.*

Our procedure will now be to express $H_1$, $H_2$, $U$, and $P$ in terms of the bending moments just defined as well as other kinematic and static quantities depending only on $x$ and $y$. The variational principle will then provide all the necessary relations among these two-dimensional quantities.

We begin by examining the following term of $H_1$:

$$\iiint \sigma_x \frac{\partial u}{\partial x}\, dxdydz = \iint \{M_x \frac{\partial}{\partial x} \int_{-h}^{h} \frac{3z}{2h^3}\, u\, dz\} dxdy.$$

The latter form provides a natural definition of the mean rotation:

$$\alpha(x,y) \equiv \frac{3}{2h^3} \int_{-h}^{h} uz\, dz.$$

One may observe that this would be satisfied identically upon introduction of the hypothesis $u(x,y,z) = z\alpha(x,y)$, in which case $\alpha$ would represent the rotation of a facet with normal

---

*Editor's note: This is a form of the *Reisser-Hellinger* variational principle. When extended in the obvious way to a general, three-dimensional, isotropic, linear elastic body, and appended by the strain-displacement relations, it may be shown to yield the equilibrium equations, stress-strain relations, and boundary conditions of the general elasticity problem. See Fung, Y. C., *Foundations of Continuum Mechanics*. Prentice-Hall, Englewood Cliffs, New Jersey, 1965, p. 299.

in the x direction. More generally, without invoking this hypothesis, we may note that $z\,dz = -d(h^2-z^2)/2$ and integrate by parts to re-express $\alpha$ as

$$\alpha = \frac{3}{2h^3}\int_{-h}^{h} \frac{h^2-z^2}{2}\,\frac{\partial u}{\partial z}\,dz.$$

In this form, $\alpha$ appears more clearly as a weighted average, over the thickness, of the local slope of the deformed facet, with weighting function $(h^2-z^2)/2$. The numerical factor preceding the integral is the inverse of the total weight

$$\int_{-h}^{h} \frac{h^2-z^2}{2}\,dz = \frac{2h^3}{3}$$

of the weighting function. Thus by introducing the mean rotations

$$\alpha = \frac{3}{2h^3}\int_{-h}^{h} \frac{h^2-z^2}{2}\,\frac{\partial u}{\partial z}\,dz, \quad \beta = \frac{3}{2h^3}\int_{-h}^{h} \frac{h^2-z^2}{2}\,\frac{\partial v}{\partial z}\,dz \tag{8.5}$$

we arrive at the two-dimensional expression

$$H_1 = \iint \left[M_x \frac{\partial \alpha}{\partial x} + M_y \frac{\partial \beta}{\partial y} + M_{xy}\left(\frac{\partial \alpha}{\partial y} + \frac{\partial \beta}{\partial x}\right)\right]dxdy.$$

In order similarly to treat $H_2$, we must first obtain the form of the stresses $\tau_{zx}$ and $\tau_{zy}$. To this end we appeal to the translational equilibrium equations

$$\frac{\partial \sigma_x}{\partial x} + \frac{\partial \tau_{yx}}{\partial y} + \frac{\partial \tau_{zx}}{\partial z} = 0, \qquad \frac{\partial \tau_{xy}}{\partial x} + \frac{\partial \sigma_y}{\partial y} + \frac{\partial \tau_{zy}}{\partial z} = 0,$$

which must be satisfied in view of their status as Euler equations connected with the unconditional variations of $u(x,y,z)$ and $v(x,y,z)$. Equations (8.2) then yield

$$\frac{\partial \tau_{zx}}{\partial z} = -\frac{3z}{2h^3}\left(\frac{\partial M_x}{\partial x} + \frac{\partial M_{yx}}{\partial y}\right), \quad \frac{\partial \tau_{zy}}{\partial z} = -\frac{3z}{2h^3}\left(\frac{\partial M_{xy}}{\partial x} + \frac{\partial M_y}{\partial y}\right)$$

and, upon integration,

$$\tau_{zx} = \frac{3}{2h^3}\frac{h^2-z^2}{2}\left(\frac{\partial M_x}{\partial x} + \frac{\partial M_{yx}}{\partial y}\right) + A(x,y),$$

$$\tau_{zy} = \frac{3}{2h^3}\frac{h^2-z^2}{2}\left(\frac{\partial M_{xy}}{\partial x} + \frac{\partial M_y}{\partial y}\right) + B(x,y).$$

Hypotheses (8.4) require that the functions $A$ and $B$ vanish. Integration over the thickness introduces the distributed shear forces

$$T_x = \int_{-h}^{h} \tau_{zx} dz, \quad T_y = \int_{-h}^{h} \tau_{zy} dz, \tag{8.6}$$

and provides the two-dimensional equilibrium equations

$$T_x = \frac{\partial M_x}{\partial x} + \frac{\partial M_{yx}}{\partial y}, \quad T_y = \frac{\partial M_{xy}}{\partial x} + \frac{\partial M_y}{\partial y}. \tag{8.7}$$

Thus, in analogy with (8.2), the transverse shear stresses take the forms

$$\tau_{zx} = \frac{3}{2h^3}\frac{h^2-z^2}{2} T_x, \quad \tau_{zy} = \frac{3}{2h^3}\frac{h^2-z^2}{2} T_y. \tag{8.8}$$

These may now be used in $H_2$ to yield the desired two-dimensional form

$$H_2 = \iint [T_x(\alpha + \frac{\partial W}{\partial x}) + T_y(\beta + \frac{\partial W}{\partial y})]dxdy,$$

where the weighted mean $W$ of the transverse displacement is given by

$$W(x,y) = \frac{3}{2h^3}\int_{-h}^{h} \frac{h^2-z^2}{2} w\, dz. \tag{8.9}$$

With the $z$-dependence of all stresses now made explicit, it is possible to integrate the complementary strain energy $U$ with respect to $z$ to obtain

$$U = \frac{1}{2E}\frac{3}{2h^3}\iint [M_x^2 + M_y^2 - 2\nu M_x M_y + 2(1+\nu)M_{xy}^2]dxdy + \frac{1}{2G}\frac{3}{5h}\iint (T_x^2 + T_y^2)dxdy.$$

We now examine the potential energy of the applied loads. With no restrictive hypothesis affecting the displacement $w(x,y,z)$, the vertical translational equilibrium equation

$$\frac{\partial \tau_{xz}}{\partial x} + \frac{\partial \tau_{yz}}{\partial y} + \frac{\partial \sigma_z}{\partial z} + Z = 0$$

should hold, where $Z$ is the body force. By using the hypothesis (8.3) and the forms (8.8), we find that $Z$ should also have a distribution parabolic in $z$, i.e.,

$$Z = q(x,y)\frac{3}{2h^3}\frac{h^2-z^2}{2}, \tag{8.10}$$

where

$$q = \int_{-h}^{h} Z\,dz \tag{8.11}$$

is the total transverse load. This transforms the equilibrium equation to

$$\frac{\partial T_x}{\partial x} + \frac{\partial T_y}{\partial y} + q = 0 \tag{8.12}$$

and puts the corresponding term of the potential energy into the two-dimensional form

$$-\iint qW\,dxdy.$$

Like the equilibrium equations, the natural (traction) boundary conditions of the three-dimensional theory follow from unconstrained displacement variations. These conditions cannot be satisfied unless the prescribed tractions take the forms

$$\bar{t}_x = \frac{3z}{2h^3} (\bar{M}_x \cos\theta + \bar{M}_{yx} \sin\theta),$$

$$\bar{t}_y = \frac{3z}{2h^3} (\bar{M}_{xy} \cos\theta + \bar{M}_y \sin\theta),$$

$$\bar{t}_z = \frac{3}{2h^3} \frac{h^2 - z^2}{2} (\bar{T}_x \cos\theta + \bar{T}_y \sin\theta),$$

where $(\cos\theta, \sin\theta, 0)$ are the direction cosines of the exterior normal to the plate along the boundary. Thus we assume these forms to obtain, and may put the corresponding term of the potential energy into the form

$$-\oint [\alpha(\bar{M}_x\cos\theta + \bar{M}_{yx}\sin\theta) + \beta(\bar{M}_{xy}\cos\theta + \bar{M}_y\sin\theta) + W(\bar{T}_x\cos\theta + \bar{T}_y\sin\theta)]ds$$

$$= -\frac{1}{2} \operatorname{Re} \oint (\alpha+i\beta)[e^{i\theta}(\bar{M}_x-\bar{M}_y-2i\bar{M}_{xy}) + e^{-i\theta}(\bar{M}_x+\bar{M}_y)]ds - \operatorname{Re}\oint W(\bar{T}_x+i\bar{T}_y)e^{-i\theta}ds.$$

By the tensor transformations

$$\alpha + i\beta = e^{i\theta}(\alpha_n + i\alpha_t), \quad M_x + M_y = M_n + M_t,$$

$$M_x - M_y - 2iM_{xy} = e^{-2i\theta}(M_n - M_t - 2iM_{nt}),$$

$$T_x + iT_y = e^{i\theta}(T_n + iT_t),$$

we obtain the form adapted to the local orientation of the boundary

$$-\frac{1}{2} \operatorname{Re} \oint (\alpha_n+i\alpha_t)(2\bar{M}_n-2i\bar{M}_{nt})ds - \operatorname{Re}\oint W(\bar{T}_n+i\bar{T}_t)ds$$

$$= -\oint (\alpha_n\bar{M}_n + \alpha_t\bar{M}_{nt} + W\bar{T}_n)ds.$$

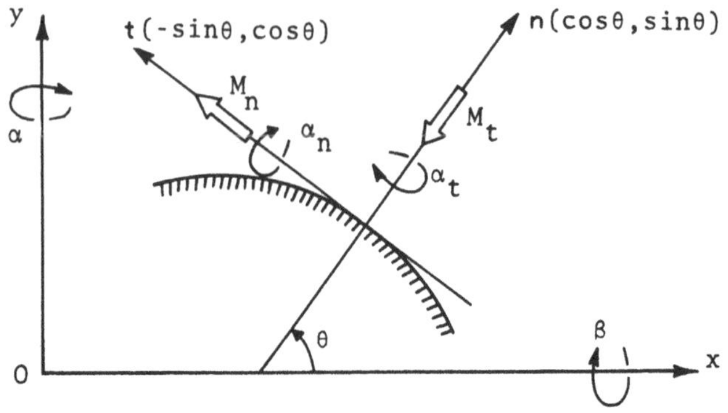


Figure 8.2

## 8.3. The Two-Dimensional Canonical Principle

By collecting the foregoing results, we obtain the two-dimensional canonical principle

$$\delta\left\{\iint [M_x \frac{\partial\alpha}{\partial x} + M_y \frac{\partial\beta}{\partial y} + M_{xy}(\frac{\partial\alpha}{\partial y} + \frac{\partial\beta}{\partial x}) + T_x(\alpha + \frac{\partial W}{\partial x}) + T_y(\beta + \frac{\partial W}{\partial y})]dx\, dy - \frac{1}{2D}\iint[\frac{M_x^2 + M_y^2 - 2\nu M_x M_y}{1-\nu^2} + \frac{2}{1-\nu}M_{xy}^2]dxdy - \frac{1}{2G}\frac{3}{5h}\iint [T_x^2 + T_y^2]dxdy - \iint qWdxdy - \oint(\alpha_n\overline{M}_n + \alpha_t\overline{M}_{nt} + W\overline{T}_n)ds\right\} = 0, \quad (8.13)$$

where the *flexural rigidity* D is defined as

$$D = \frac{2}{3}\frac{Eh^3}{1-\nu^2} . \quad (8.14)$$

In this principle the three kinematic variables $(\alpha,\beta,W)$ and the five stress resultants $(M_x, M_y, M_{xy}, T_x, T_y)$ may be varied independently and arbitrarily, provided that $\alpha$, $\beta$, and W take the proper values on parts of the boundary where displacements are specified, and do not vary there. As may be verified easily, the Euler equations connected with the variations of $\alpha$, $\beta$, and $W$ are respectively the rotational equilibrium equations (8.7) and the vertical translational equilibrium equation (8.12) for a small portion $2h\,dxdy$ of the plate. After an application of Gauss' theorem and a tensor transformation analogous to the foregoing, the boundary terms generated by the variation of the surface integrals become

$$\oint (M_n \delta\alpha_n + M_{nt}\delta\alpha_t + T_n \delta W)ds.$$

When combined with the last term of (8.13) these generate the natural boundary conditions

$$M_n = \overline{M}_n, \quad M_{nt} = \overline{M}_{nt}, \quad T_n = \overline{T}_n. \tag{8.15}$$

The variations in $M_x$, $M_{xy}$, and $M_y$ supply the relations

$$\frac{\partial\alpha}{\partial x} = \frac{1}{D(1-\nu^2)}(M_x - \nu M_y), \quad \frac{\partial\beta}{\partial y} = \frac{1}{D(1-\nu^2)}(M_y - \nu M_x),$$

$$\frac{\partial\alpha}{\partial y} + \frac{\partial\beta}{\partial x} = \frac{1}{D}\frac{2}{1-\nu}M_{xy},$$

which in inverse form become the moment-rotation relations

$$M_x = D\left(\frac{\partial\alpha}{\partial x} + \nu\frac{\partial\beta}{\partial y}\right), \quad M_y = D\left(\frac{\partial\beta}{\partial y} + \nu\frac{\partial\alpha}{\partial x}\right),$$

$$M_{xy} = D\frac{1-\nu}{2}\left(\frac{\partial\beta}{\partial x} + \frac{\partial\alpha}{\partial y}\right). \tag{8.16}$$

(Note how the inverse form justifies the definition of $D$.) Finally, the variations of $T_x$ and $T_y$ supply the shear deformation relations

$$T_x = 2nGh\left(\alpha + \frac{\partial W}{\partial x}\right), \quad T_y = 2nGh\left(\beta + \frac{\partial W}{\partial y}\right) \tag{8.17}$$

where

$$n = 5/6. \tag{8.18}$$

Let us now express the generalized stress-strain laws (8.16), (8.17) at the boundaries in the natural directions. By starting with the equivalent form of (8.16)

$$M_x + M_y = (1+\nu)D\,\mathrm{Re}[(\frac{\partial}{\partial x} - i\,\frac{\partial}{\partial y})(\alpha + i\beta)],$$

$$M_x - M_y - 2iM_{xy} = (1-\nu)D\,(\frac{\partial}{\partial x} - i\,\frac{\partial}{\partial y})(\alpha - i\beta),$$

and using the preceding tensor transformations along with the relation

$$\frac{\partial}{\partial x} - i\,\frac{\partial}{\partial y} = e^{-i\theta}(\frac{\partial}{\partial n} - i\,\frac{\partial}{\partial s}), \qquad (8.19)$$

we obtain

$$M_n+M_t = (1+\nu)D\,\mathrm{Re}[e^{-i\theta}(\frac{\partial}{\partial n} - i\,\frac{\partial}{\partial s})e^{i\theta}(\alpha_n+i\alpha_t)],$$

$$M_n-M_t-2iM_{nt} = (1-\nu)D\,e^{i\theta}(\frac{\partial}{\partial n} - i\,\frac{\partial}{\partial s})e^{-i\theta}(\alpha_n-i\alpha_t).$$

The curvature of the boundary is $\dot{\theta} \equiv \partial\theta/\partial s$, and $\partial\theta/\partial n = 0$, so we have

$$M_n+M_t = (1+\nu)D\,\mathrm{Re}[(\frac{\partial}{\partial n} - i\,\frac{\partial}{\partial s})(\alpha_n+i\alpha_t) + \dot{\theta}(\alpha_n+i\alpha_t)],$$

$$M_n-M_t-2iM_{nt} = (1-\nu)D\,[(\frac{\partial}{\partial n} - i\,\frac{\partial}{\partial s})(\alpha_n - i\alpha_t) - \dot{\theta}(\alpha_n-i\alpha_t)],$$

whence, in particular,

$$M_n = D[\frac{\partial\alpha_n}{\partial n} + \nu\,(\frac{\partial\alpha_t}{\partial s} + \dot{\theta}\alpha_n)], \qquad (8.20)$$

$$M_{nt} = D\,\frac{1-\nu}{2}\,[\frac{\partial\alpha_t}{\partial n} + \frac{\partial\alpha_n}{\partial s} - \dot{\theta}\alpha_t)]. \qquad (8.21)$$

A similar calculation yields

$$T_n = 2nGh\,[\alpha_n + \frac{\partial W}{\partial n}]. \qquad (8.22)$$

At each point of the boundary eight different triplets of boundary conditions are possible according to the three independent alternatives

$$\alpha_n = \overline{\alpha}_n \quad \text{or} \quad M_n = \overline{M}_n,$$
$$\alpha_t = \overline{\alpha}_t \quad \text{or} \quad M_{nt} = \overline{M}_{nt}, \tag{8.23}$$
$$W = \overline{W} \quad \text{or} \quad T_n = \overline{T}_n.$$

As before, an overbar denotes a prescribed quantity. Among the homogeneous combinations occurring most frequently in practice are those for a *free edge,* $M_n = M_{nt} = T_n = 0$, and for a *clamped edge,* $\alpha_n = \alpha_{nt} = W = 0$. A *simply supported edge* is approximated by either $M_n = M_{nt} = W = 0$ or $M_n = \alpha_t = W = 0$. In the presence of certain symmetries, other homogeneous conditions may be obtained. Thus if the geometry and loading of plate are symmetric with respect to an axis $t$ in the middle surface, then along that axis $\alpha_n = M_{nt} = T_n = 0$.

## 8.4. Further Connections Between the Two- and Three-Dimensional Theories

We have adopted approximations only for the forms of $\sigma_x$, $\sigma_y$, $\sigma_z$, and $\tau_{xy}$. The Euler equations of the three-dimensional canonical principle (8.2) corresponding to the other stresses and the displacements must therefore still be satisfied. The three-dimensional equilibrium equations, which correspond to variations of displacements, have already been invoked. The stress-strain relations

$$\tau_{zx} = G\left(\frac{\partial u}{\partial z} + \frac{\partial w}{\partial x}\right), \qquad \tau_{zy} = G\left(\frac{\partial v}{\partial z} + \frac{\partial w}{\partial y}\right) \tag{8.24}$$

are the Euler equations connected with variations of $\tau_{zx}$ and $\tau_{zy}$, and therefore must be satisfied. By virtue of (8.8) and the definitions (8.5) and (8.9), when multiplied by $3(h^2-z^2)/4h^3$ and integrated across the thickness, these equations develop again into equations (8.17), with which, of course, they cannot be in contradiction. They thus contribute nothing further to the two-dimensional theory, and it is complete. They can still supply supplementary information about the displacement field. For example, if one adopts the approximative hypothesis

$$w(x,y,z) = W(x,y),$$

which is compatible with the definition (8.9) which it reduces to an identity, one can satisfy exactly the preceding equations (8.24) and the conditions of symmetry of the problem with the field

$$u = \frac{3}{4Gh^3}\left(h^2 z - \frac{1}{3} z^3\right)T_x - z\frac{\partial W}{\partial x},$$

$$v = \frac{3}{4Gh^3}\left(h^2 z - \frac{1}{3} z^3\right)T_y - z\frac{\partial W}{\partial y}.$$

The only approximations still remaining in the present two-dimensional theory are those affecting the stress-strain equations

$$E\frac{\partial u}{\partial x} = \sigma_x - \nu\sigma_y, \quad E\frac{\partial v}{\partial y} = \sigma_y - \nu\sigma_x, \quad G\left(\frac{\partial u}{\partial y} + \frac{\partial v}{\partial x}\right) = \tau_{xy},$$

which are the Euler equations in the three-dimensional theory for variations of $(\sigma_x, \sigma_y, \tau_{xy})$. In general, they are satisfied only in the weighted mean. This is easily verified by multiplying them by $z$ and integrating over the thickness to recover equations (8.16).

## 8.5. Other Types of Approximations

The fundamental approximative assumption in *Hencky's* theory is that the displacements take the forms

$$u = z\alpha(x,y), \quad v = z\beta(x,y), \quad w = W(x,y). \tag{8.25}$$

This may be viewed as a restrictive interpretation of the definitions (8.5) and (8.9). The quantities $(\alpha,\beta,W)$, functions of only two variables, are then chosen according to the canonical variational principle. The three-dimensional version of it requires the strain-displacement and stress-strain laws to be satisfied. The strains are therefore

$$\varepsilon_x = z\frac{\partial\alpha}{\partial x},\quad \varepsilon_y = z\frac{\partial\beta}{\partial y},\quad \gamma_{xy} = z\left(\frac{\partial\alpha}{\partial y} + \frac{\partial\beta}{\partial x}\right),$$

$$\varepsilon_z = 0\quad,\quad \gamma_{zx} = \frac{\partial W}{\partial x} + \alpha,\quad \gamma_{zy} = \frac{\partial W}{\partial y} + \beta.$$

These expressions and the stress-strain laws yield, in particular,

$$\sigma_z = \nu(\sigma_x + \sigma_y),$$

with a fundamental difference from (8.3). Also, the shear stress distributions

$$\tau_{zx} = G\left(\frac{\partial W}{\partial x} + \alpha\right),\quad \tau_{zy} = G\left(\frac{\partial W}{\partial y} + \beta\right) \tag{8.26}$$

are constant rather than parabolic as in (8.8), and the boundary conditions (8.4) no longer hold.

With the stress-strain and strain-displacement laws in force, the canonical principle reduces to that of minimum potential energy, with only the displacements subject to variation. As before, the prescribed $z$-dependence permits integration over the thickness, reducing the problem from three to two dimensions. With the stress resultants defined by (8.1) and (8.6), the equations of the corresponding two-dimensional theory are formally identical with those of the preceding theory, with $D$ replaced by the larger value $2Eh^3/3(1-\nu^2)^2$, and the transverse shear coefficient $n$ in (8.17) increased from $5/6$ to $1$.

In view of the hypothesis $w = W(x,y)$, the potential of the transverse loads may be written

$$-\iiint Z\,w\,dxdydz - \iint [(t_z w)_{z=h} + (t_z w)_{z=-h}]dxdy$$
$$= - \iint q\,W\,dxdy$$

where $q(x,y)$ depends only on the sum

$$q = \int_{-h}^{h} Z\,dz + \sigma_z(x,y,h) - \sigma_z(x,y,-h).$$

The theory presented in Sections (8.1)-(8.4) used fundamental assumptions on the stresses, while Hencky approximated the displacements. *Reissner*, in contrast, simultaneously hypothesized the forms (8.25) for the displacements and (8.2)-(8.3) for the stresses. The forms of the shear stresses $\tau_{zx}$ and $\tau_{zy}$ then follow from (8.24), the only Euler equations of the three-dimensional theory which must still hold, and are found to be constant through the thickness. This yields the value unity for the transverse shear coefficient $n$. Integration of the canonical principle through the thickness, and variation of the two-dimensional field quantities, yields the same equations as in the other cases, with the same value for the flexural rigidity $D$ as in the basic theory.

Henceforth we shall discuss together the equations produced by the preceding theories, given their formal identity. The flexural rigidity will always be denoted by $D$, and the shear coefficient of (8.17) by $n$.

## 8.6. Kirchhoff's Hypothesis

An additional approximative hypothesis, originally introduced by *Kirchhoff*, is that each fiber perpendicular to the middle surface in the initial configuration remains perpendicular to it during the deformation. In the context of either Hencky's or Riessner's theory, where the fundamental assumptions involve displacements, this may be expressed as

$$\alpha + \partial W/\partial x = 0, \qquad \beta + \partial W/\partial y = 0, \tag{8.27}$$

which in turn implies that the transverse shear strains vanish identically. Conditions (8.27) may also be incorporated into the basic theory of Sections (8.1)-(8.4), wherein they express no more than a perpendicularity in the sense of a weighted mean. In either context, they allow the rotations to be replaced by gradients of a single transverse displacement, and reduce the force-deformation relations to the *moment-curvature* relations

$$M_x = -D\left(\frac{\partial^2 W}{\partial x^2} + \nu\frac{\partial^2 W}{\partial y^2}\right), \qquad M_y = -D\left(\frac{\partial^2 W}{\partial y^2} + \nu\frac{\partial^2 W}{\partial x^2}\right),$$

$$M_{xy} = -D(1-\nu)\frac{\partial^2 W}{\partial x \partial y}. \tag{8.28}$$

On the edges the conditions of perpendicularity become

$$\alpha_n + \frac{\partial W}{\partial n} = 0, \qquad \alpha_t + \frac{\partial W}{\partial s} = 0. \tag{8.29}$$

This transforms (8.20) into the form

$$M_n = -D\left[\frac{\partial^2 W}{\partial n^2} + \nu\left(\frac{\partial^2 W}{\partial s^2} + \dot{\theta}\frac{\partial W}{\partial n}\right)\right], \tag{8.30}$$

and (8.21), initially, into

$$M_{nt} = -D\frac{1-\nu}{2}\left[\frac{\partial}{\partial n}\frac{\partial W}{\partial s} + \frac{\partial}{\partial s}\frac{\partial W}{\partial n} - \dot{\theta}\frac{\partial W}{\partial s}\right].$$

One may simplify the latter formula, and obtain a form for Laplace's operator which will be useful later, by observing that

$$\nabla^2 = \partial\bar{\partial} = e^{-i\theta}\left(\frac{\partial}{\partial n} - i\,\frac{\partial}{\partial s}\right)\left[e^{i\theta}\left(\frac{\partial}{\partial n} + i\,\frac{\partial}{\partial s}\right)\right]$$

is a real operator. Annulling the imaginary part yields the relation

$$\frac{\partial}{\partial n}\frac{\partial}{\partial s} - \frac{\partial}{\partial s}\frac{\partial}{\partial n} + \dot{\theta}\,\frac{\partial}{\partial s} = 0,$$

which transforms the formula for $M_{nt}$ into

$$M_{nt} = -D(1-\nu)\left[\frac{\partial}{\partial s}\frac{\partial W}{\partial n} - \dot{\theta}\,\frac{\partial W}{\partial s}\right], \tag{8.31}$$

and that for Laplace's operator into

$$\nabla^2 = \frac{\partial^2}{\partial n^2} + \frac{\partial^2}{\partial s^2} + \dot{\theta}\,\frac{\partial}{\partial n}\,. \tag{8.32}$$

The shear force-shear deformation relations (8.17) disappear completely in Kirchhoff's theory, because they came from variational relations connected with the transverse shear forces, and these will no longer appear in the canonical principle. Indeed, $H_2$ disappears by virtue of (8.27), while the integrand of $H_1$ becomes

$$-\left(M_x\,\frac{\partial^2 W}{\partial x^2} + M_y\,\frac{\partial^2 W}{\partial y^2} + 2M_{xy}\,\frac{\partial^2 W}{\partial x\partial y}\right).$$

By virtue of Clapeyron's interior theorem, if we divide the latter expression by 2 it becomes the strain energy density, and in terms only of moments takes the form

$$\frac{1}{2D(1-\nu^2)}\left[(M_x + M_y)^2 + 2(1+\nu)(M_{xy}^2 - M_xM_y)\right].$$

Although the shear strains vanish, the shear forces do not; they may in fact be calculated from the equilibrium equations (8.7), but are reduced to the status of auxiliary variables. The variational equations (8.17) also disappear because the canonical principle now generates only the equilibrium equation connected with the variations of $W$

$$\frac{\partial^2 M_x}{\partial x^2} + 2\frac{\partial^2 M_{xy}}{\partial x \partial y} + \frac{\partial^2 M_y}{\partial y^2} + q = 0. \tag{8.33}$$

This is the old equation (8.12) in which the shear forces have been eliminated by using (8.7).

These results may also be interpreted in a very useful and somewhat less artificial manner by regarding the plate as formed of a transversely isotropic material, obtained by letting the shear modulus connected with transverse shear become infinite. This amounts to allowing the coefficient $n$ to approach infinity. Equations (8.27) follow by dividing equations (8.17) by $n$ and passing to the limit.

In view of the foregoing remarks, Kirchhoff's hypothesis transforms the canonical variational principle into the form

$$\begin{aligned}\delta\Big\{-\iint\Big[M_x\frac{\partial^2 W}{\partial x^2} + M_y\frac{\partial^2 W}{\partial y^2} + 2M_{xy}\frac{\partial^2 W}{\partial x\partial y}\Big]dxdy \\ - \frac{1}{2D(1-\nu^2)}\iint\big[(M_x+M_y)^2 + 2(1+\nu)(M_{xy}^2 - M_xM_y)\big]dxdy \\ - \iint qWdxdy + \oint\Big(\overline{M}_n\frac{\partial W}{\partial n} + \overline{M}_{nt}\frac{\partial W}{\partial s} - \overline{T}_n W\Big)ds\Big\} = 0.\end{aligned} \tag{8.34}$$

The variational equations of this principle are (8.33) for $\delta W$ and (8.28) for the variations in the moments. The natural boundary conditions require special attention.

## 8.7. Boundary Conditions in Kirchhoff's Theory

The boundary terms of the principle (8.34) initially appear in the form

$$\oint [-(\cos\theta\ M_x + \sin\theta\ M_{yx})\delta\frac{\partial W}{\partial x} - (\cos\theta\ M_{xy} + \sin\theta\ M_y)\delta\frac{\partial W}{\partial y}$$

$$+ \delta W \cos\theta\left(\frac{\partial M_x}{\partial x} + \frac{\partial M_{yx}}{\partial y}\right) + \delta W \sin\theta\left(\frac{\partial M_{xy}}{\partial x} + \frac{\partial M_y}{\partial y}\right)$$

$$+ (\bar{M}_n\ \delta\frac{\partial W}{\partial n} + \bar{M}_{nt}\ \delta\frac{\partial W}{\partial s} - \bar{T}_n\ \delta W)]ds = 0.$$

When the derivatives of the variation of the deflection $W$ are expressed in the natural directions of the edge with the help of (8.19), and the shear forces are introduced according to the equilibrium equations (8.7), then tensor transformations of the moments yield

$$\oint (\bar{M}_n - M_n)\delta\frac{\partial W}{\partial n}\,ds + (\bar{M}_{nt} - M_{nt})\frac{\partial}{\partial s}\,\delta W\,ds + (T_n - \bar{T}_n)\delta W\,ds = 0.$$

If the variation $\delta W$ is known along the edge, its derivative $\partial(\delta W)/\partial s$ in the direction of the edge is also known and is not independent. It is thus appropriate to integrate the second term by parts. The formal result is

$$[(\bar{M}_{nt} - M_{nt})\delta W\oint + \oint (\bar{M}_n - M_n)\delta\frac{\partial W}{\partial n}\,ds + (K_n - \bar{K}_n)\delta W\,ds = 0,$$

with a modified shear force, called *Kirchhoff's shear force*, defined by

$$K_n = T_n + \frac{\partial M_{nt}}{\partial s}. \tag{8.35}$$

If the edge is a regular curve, without angular points, and the applied twisting moment $\bar{M}_{nt}$ is continuous, then the cyclic term vanishes:

$$[(\bar{M}_{nt} - M_{nt})\delta W\oint = 0.$$

More generally, however, the contour will comprise regular arcs separated by angular points. By decomposing the contour into its regular arcs, we may express the term in question as

$$\sum_{\text{arcs}} [(\overline{M}_{nt} - M_{nt})_{s=s_j-0}\, \delta W_j - (\overline{M}_{nt} - M_{nt})_{s=s_k+0}\, \delta W_k]$$

where $s_j$ and $s_k$ are the terminal and initial points of each arc. By grouping together all terms from the same angular point, the sum becomes

$$\sum_i (\overline{Z}_i - \Delta_i M_{nt})\delta W_i$$

where

$$\overline{Z}_i = \overline{M}_{nt}(s_i + 0) - \overline{M}_{nt}(s_i - 0)$$

is a *corner force* and

$$\Delta_i M_{nt} = M_{nt}(s_i + 0) - M_{nt}(s_i - 0) \qquad (8.36)$$

is the discontinuity in the internal twisting moment caused by the change in orientation of the contour at the angular point.

The following interpretation is due to *Kelvin* and *Tait*.

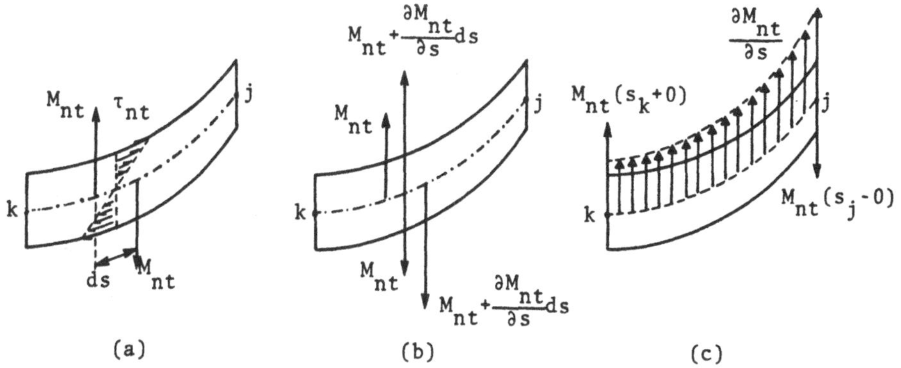


Figure 8.3

On a strip of a regular arc of length ds the shear stresses $\tau_{nt}$ form a couple of strength $M_{nt}ds$. In the sense of statics this couple can always be replaced by an equivalent couple formed by two vertical forces of strength $M_{nt}$ separated by a distance of ds (Figure 8.3a). This equivalence may also be demonstrated by noting that the virtual work of the two forces is

$$M_{nt}W - M_{nt}(W + \frac{\partial W}{\partial s} ds) = -M_{nt} \frac{\partial W}{\partial s} ds$$

and, in view of the second of (8.29), that this virtual work is also $M_{nt}\ \alpha_t$ ds, that of the shear stresses $\tau_{nt}$. The successive replacement by couples of vertical force has for each regular arc the final effect (Figures 8.3a,b) of localized vertical forces $M_{nt}(s_k + 0)$ and $M_{nt}(s_j - 0)$ at the ends of the arc and a *replacement shear force* of intensity $\partial M_{nt}/\partial s$.

The boundary conditions of static type in Kirchhoff's theory are therefore

$$M_n = \overline{M}_n, \tag{8.37}$$

generated in the canonical principle by the variations $\delta \frac{\partial W}{\partial n}$ at the boundaries;

$$K_n = \overline{K}_n, \tag{8.38}$$

caused by variations $\delta W$ at the boundaries; and

$$\Delta_i\ M_{nt} = \overline{Z}_i, \tag{8.39}$$

due to local variations in the deflection at each angular point. In principle, then, the loads imposable on the edges are formed by bending moments, shear forces, and concentrated forces at the angular points.

The conditions of kinematic type which may be imposed apply only to the deflection or its normal slope.

The table of alternatives for each edge is the following:

$$\frac{\partial W}{\partial n} = -\bar{\alpha}_n \quad \text{or} \quad M_n = \bar{M}_n,$$
$$W = \bar{W} \quad \text{or} \quad \begin{cases} K_n = \bar{K}_n \\ \Delta_i\, M_{nt} = \bar{Z}_i \end{cases} . \tag{8.40}$$

Kirchhoff's hypothesis thus has the effect of restricting the admissible boundary conditions; in particular it is no longer possible to impose separately, as in the preceding theories, each of the resultants of the surface tractions.

The most common homogeneous conditions are $M_n = K_n = 0$ for a free edge, $M_n = W = 0$ for a simply supported edge, and $W = \partial W/\partial n = 0$ for a clamped edge. If there is symmetry with respect to an axis, then $\partial W/\partial n = K_n = 0$.

A useful expression for the shear force $T_n$ may be found by starting with the equilibrium equations to obtain

$$2(T_x + iT_y) = \partial(M_x - M_y + 2iM_{xy}) + \bar{\partial}(M_x + M_y).$$

By tensor transformation this becomes

$$\begin{aligned} 2(T_n + iT_t) &= e^{-i\theta}[e^{-i\theta}(\frac{\partial}{\partial n} - i\,\frac{\partial}{\partial s})e^{2i\theta}(M_n - M_t + 2iM_{nt}) \\ &\qquad + e^{i\theta}(\frac{\partial}{\partial n} + i\,\frac{\partial}{\partial s})(M_n + M_t)] \\ &= (\frac{\partial}{\partial n} - i\,\frac{\partial}{\partial s})(M_n - M_t + 2iM_{nt}) + 2\dot{\theta}(M_n - M_t + 2iM_{nt}) \\ &\qquad + (\frac{\partial}{\partial n} + i\,\frac{\partial}{\partial s})(M_n + M_t) \end{aligned}$$

and, in particular,

$$T_n = \partial M_n/\partial n + \partial M_{nt}/\partial s + \dot{\theta}(M_n - M_t). \tag{8.41}$$

By adding the replacement shear force this yields

$$K_n = \partial M_n/\partial n + 2\ \partial M_{nt}/\partial s + \dot{\theta}(M_n - M_t). \tag{8.42}$$

From (8.7) and (8.28) we also have

$$T_x = -D \frac{\partial}{\partial x} \nabla^2 W, \quad T_y = -D \frac{\partial}{\partial y} \nabla^2 W, \tag{8.43}$$

and therefore

$$T_n = -D \frac{\partial}{\partial n} \nabla^2 W. \tag{8.44}$$

If we now replace Laplace's operator by (8.32) and use (8.31) we have

$$K_n = -D \frac{\partial}{\partial n} \left(\frac{\partial^2 W}{\partial n^2} + \dot{\theta} \frac{\partial W}{\partial n} + \frac{\partial^2 W}{\partial s^2}\right) - D(1-\nu)\frac{\partial}{\partial s}\left(\frac{\partial}{\partial s} \frac{\partial W}{\partial n} - \dot{\theta} \frac{\partial W}{\partial s}\right). \tag{8.45}$$

## 8.8. Kirchhoff's Variational Principle

By using the moment-curvature relations (8.28) and the definition (8.35) of Kirchhoff's shear force, the variational principle (8.34) may be recast in the form

$$\delta\left\{\frac{1}{2} \iint D\{(\nabla^2 W)^2 + 2(1-\nu)\ [(\frac{\partial^2 W}{\partial x \partial y})^2 - \frac{\partial^2 W}{\partial x^2} \frac{\partial^2 W}{\partial y^2}]\} dxdy\right.$$

$$\left. - \iint qW\ dxdy - \sum_i W_i \overline{Z}_i - \oint [\overline{K}_n W - \overline{M}_n \frac{\partial W}{\partial n}] ds\right\} = 0. \tag{8.46}$$

In the strain energy the term depending on Poisson's ratio is a divergence and may therefore by transferred to the boundaries, i.e.,

$$\left(\frac{\partial^2 W}{\partial x \partial y}\right)^2 - \frac{\partial^2 W}{\partial x^2} \frac{\partial^2 W}{\partial y^2} = \frac{\partial A}{\partial x} + \frac{\partial B}{\partial y}$$

with, for example,

$$A = \frac{\partial W}{\partial y} \cdot \frac{\partial^2 W}{\partial x \partial y}, \quad B = -\frac{\partial W}{\partial y} \cdot \frac{\partial^2 W}{\partial x^2}.$$

Therefore

$$(1-\nu)D \iint \left[\left(\frac{\partial^2 W}{\partial x \partial y}\right)^2 - \frac{\partial^2 W}{\partial x^2}\frac{\partial^2 W}{\partial y^2}\right] dxdy$$
$$= (1-\nu)D \oint \frac{\partial W}{\partial y}\, d\, \frac{\partial W}{\partial x} . \tag{8.47}$$

This can also be given the following, more symmetric form, allowing ultimate passage to the local directions:

$$\frac{1-\nu}{2} D \oint \frac{\partial W}{\partial y}\, d\, \frac{\partial W}{\partial x} - \frac{\partial W}{\partial x}\, d\, \frac{\partial W}{\partial y}$$
$$= \frac{1-\nu}{2} D \oint \frac{\partial W}{\partial s}\, d\, \frac{\partial W}{\partial n} - \frac{\partial W}{\partial n}\, d\, \frac{\partial W}{\partial s} - \left[\left(\frac{\partial W}{\partial n}\right)^2 + \left(\frac{\partial W}{\partial s}\right)^2\right] d\theta. \tag{8.48}$$

It follows that this contribution vanishes in the case of a completely clamped edge.

By performing the variation indicated in (8.46) and using (8.47), the Euler equation is found to be

$$\nabla^2\nabla^2 W = \frac{1}{D}\, q, \tag{8.49}$$

which is also known as *Sophie Germain's* equation. We also obtain

$$-D\nabla^2 W + (1-\nu)D\left(\frac{\partial^2 W}{\partial s^2} + \dot{\theta}\frac{\partial W}{\partial n}\right) = \overline{M}_n \tag{8.50}$$

as a natural condition following from the annihilation of the coefficient of the variation $\delta(\partial W/\partial n)$ in the curvilinear integral. This is equation (8.37) as modified by (8.30) and (8.32).

Similarly, the annihilation of the coefficient of the variation $\delta W$ in the curvilinear integral yields

$$-D\frac{\partial}{\partial n}\nabla^2 W - D(1-\nu)\frac{\partial}{\partial s}\left(\frac{\partial^2 W}{\partial s \partial n} - \dot{\theta}\frac{\partial W}{\partial s}\right) = \overline{K}_n, \tag{8.51}$$

which is equivalent to (8.38) as modified by (8.44) and (8.31). And by varying the deflection at the angular points, we obtain the equivalent of (8.39) as modified by (8.31)

$$-D\ (1-\nu)\Delta_i\ \left(\frac{\partial^2 W}{\partial s \partial n} - \dot{\theta}\ \frac{\partial W}{\partial s}\right) = \overline{Z}_i. \tag{8.52}$$

## 8.9. Structure of the Solution of the Equations of Plates of Moderate Thickness

Sophie Germain's equation governing the transverse displacement in Kirchhoff's theory for isotropic plates suggests the use of biharmonic functions and hence of analytic functions of the complex variable $\zeta = x + iy$. One might wonder whether the inclusion of the shear deformations of the more elaborate theories could alter the basic character of the solutions. In fact, as we shall see in the following development, one part (the principal part) of the solution will continue to depend on analytic functions, while the remainder will be controlled by another elliptic operator. The latter will permit a rapid adjustment of the stresses and strains, over distances of the order of the thickness of the plate, so as to meet the three boundary conditions required by the more elaborate theories. At issue, then, is an "edge effect" and we now seek its mathematical origin and physical interpretation.

We begin by eliminating the moments and shear forces from the equations for plates of moderate thickness:

$$\frac{\partial M_x}{\partial x} + \frac{\partial M_{yx}}{\partial y} = T_x, \qquad \frac{\partial M_{xy}}{\partial x} + \frac{\partial M_y}{\partial y} = T_y \tag{e.1}$$

$$\frac{\partial T_x}{\partial x} + \frac{\partial T_y}{\partial y} = -q \tag{e.2}$$

$$M_x = D\left(\frac{\partial \alpha}{\partial x} + \nu\ \frac{\partial \beta}{\partial y}\right), \quad M_y = D\left(\frac{\partial \beta}{\partial y} + \nu\ \frac{\partial \alpha}{\partial x}\right), \quad M_{xy} = D\ \frac{1-\nu}{2}\ \left(\frac{\partial \beta}{\partial x} + \frac{\partial \alpha}{\partial y}\right) \tag{c.1}$$

$$T_x = 2\ Ghn\left(\alpha + \frac{\partial W}{\partial x}\right), \quad T_y = 2\ Ghn\left(\beta + \frac{\partial W}{\partial y}\right). \tag{c.2}$$

By eliminating the shear forces between (e.1) and (e.2) and then substituting the moments taken from (c.1) we obtain

$$\nabla^2\left(\frac{\partial\alpha}{\partial x} + \frac{\partial\beta}{\partial y}\right) = -\frac{q}{D}\,. \tag{n.1}$$

Substitution of (c.2) in (e.2) yields

$$\frac{\partial\alpha}{\partial x} + \frac{\partial\beta}{\partial y} + \nabla^2 W = -\frac{q}{2Ghn}\,. \tag{n.2}$$

A third equation comes from eliminating $W$ from the equations (c.2), substituting the shear forces from (e.1), then the moments from (c.1). The result can be expressed in terms of the quantity

$$\Omega \equiv \frac{1}{2}\left(\frac{\partial\beta}{\partial x} - \frac{\partial\alpha}{\partial y}\right) \tag{t.1}$$

in the form of the equation

$$4\,Ghn\,\Omega = (1-\nu)\,D\nabla^2\Omega$$

or, after substitution for $D$ from (8.14),

$$\nabla^2\Omega = \frac{3n}{h^2}\,\Omega. \tag{n.3}$$

The physical meaning of the quantity $\Omega$ is simple. Since $\omega = \frac{1}{2}(\partial v/\partial x - \partial u/\partial y)$ is the material rotation about an axis parallel to $Oz$, the local twist of a fiber perpendicular to the middle surface is

$$\frac{\partial\omega}{\partial z} = \frac{1}{2}\frac{\partial}{\partial x}\left(\frac{\partial v}{\partial z}\right) - \frac{1}{2}\frac{\partial}{\partial y}\left(\frac{\partial u}{\partial z}\right).$$

If this equation is multiplied by the parabolic weighting function $3(h^2-z^2)/4h^3$ and integrated through the thickness, then by the definitions (8.5) the right hand side becomes that of (t.1), and we arrive at the interpretation of

$$\Omega = \frac{3}{2h^3} \int_{-h}^{h} \frac{h^2 - z^2}{2} \frac{\partial \omega}{\partial z} \, dz$$

as the weighted mean of the twist of a fiber. A similar interpretation is possible in the context of the theories of Hencky or Reissner, in which the twist of the fiber is uniform.

This twist, when it occurs, causes important variations over distances of the order of magnitude of the thickness of the plate, since a typical solution of the equation (n.3) is

$$\Omega = C \exp(-\lambda \frac{x}{h} - \mu \frac{y}{h}), \qquad \lambda^2 + \mu^2 = 3n.$$

On the other hand, Navier's equations (n.1) and (n.2) yield solutions of the same order of regularity as the distribution $q$ of transverse load. In connection with Sophie Germain's equation, it is interesting to note that (n.1) and (n.2) combine to yield

$$\nabla^2 \nabla^2 W = \frac{q}{D} - \frac{1}{2Ghn} \nabla^2 q \tag{n.4}$$

which shows that the deflection is not altered by the edge effect.

In order to fix the structure of the general solution, we first examine the situation when the plate has no transverse load, so that $q = 0$. The equilibrium equation (e.2) then becomes homogeneous and may be satisfied by introducing a potential function $\psi$ such that

$$T_x = -(1-\nu) D \frac{\partial \psi}{\partial y}, \qquad T_y = (1-\nu) D \frac{\partial \psi}{\partial x} . \tag{e.3}$$

Substitution of (e.3) and the moments from (c.1) into (e.1) yields

$$\frac{\partial}{\partial x} \left(\frac{\partial \alpha}{\partial x} + \frac{\partial \beta}{\partial y}\right) = \frac{\partial}{\partial y} \{(1-\nu)(\Omega - \psi)\},$$

$$\frac{\partial}{\partial y} \left(\frac{\partial \alpha}{\partial x} + \frac{\partial \beta}{\partial y}\right) = -\frac{\partial}{\partial x} \{(1-\nu)(\Omega - \psi)\},$$

which are in the form of the Cauchy-Riemann equations. We can therefore find an analytic function $A(\zeta)$ such that

$$\frac{\partial\alpha}{\partial x} + \frac{\partial\beta}{\partial y} = 2\left(\frac{dA}{d\zeta} + \overline{\frac{dA}{d\zeta}}\right) \qquad (r.1)$$

and

$$(1-\nu)(\Omega-\psi) = \frac{2}{i}\left(\frac{dA}{d\zeta} - \overline{\frac{dA}{d\zeta}}\right),$$

or, by rearranging the latter,

$$\psi = \Omega + \frac{2i}{1-\nu}\left(\frac{dA}{d\zeta} - \overline{\frac{dA}{d\zeta}}\right). \qquad (r.2)$$

Also, elimination of $W$ between equations (c.2) and substitution of (e.3) yields $(1-\nu)D\nabla^2\psi = 4\,Ghn\Omega$ or, with $D$ replaced by (8.14),

$$\Omega = \frac{h^2}{3n}\nabla^2\psi. \qquad (f)$$

Along with the definition (t.1) of the twist of a fiber, the results (r.1) and (f) now allow us to find an expression for the complex operator

$$\partial = \frac{\partial}{\partial x} - i\frac{\partial}{\partial y}$$

applied to $\alpha + i\beta$, viz.,

$$\partial(\alpha+i\beta) = \frac{\partial\alpha}{\partial x} + \frac{\partial\beta}{\partial y} + 2i\Omega = 2\left(\frac{dA}{d\zeta} + \overline{\frac{dA}{d\zeta}}\right) + \frac{2ih^2}{3n}\partial\bar{\partial}\psi.$$

By using the lemmas on integration developed in Section 7.1, we obtain

$$\alpha + i\beta = A + \zeta\,\overline{\frac{dA}{d\zeta}} + \frac{2ih^2}{3n}\bar{\partial}\psi + \overline{\frac{dB}{d\zeta}} \qquad (d.1)$$

where $B(\zeta)$ is a new analytic function.

To find the transverse displacement we use equations (c.2) and (e.3) to form the combination

$$\partial W = -(\alpha-i\beta) - \frac{2ih^2}{3n}\partial\psi. \qquad (d.2)$$

By substituting the complex conjugate of (d.1) into (d.2) we obtain

$$\partial W = -\bar{A} - \bar{\zeta}\,\frac{dA}{d\zeta} - \frac{dB}{d\zeta}\,,$$

the general real-valued integral of which is

$$W = -\frac{1}{2}\,(\zeta\bar{A} + \bar{\zeta}A + B + \bar{B}) = -\mathrm{Re}[\bar{\zeta}A + B]\,. \qquad (d.3)$$

As might have been expected, the deflection has the typical structure of a biharmonic function, while the rotations $\alpha$ and $\beta$ have an analytic part and another depending on the twist of the fibers. In order to bring this out more clearly in (d.1), we apply the operator $\partial$ to the equation (r.2) to get

$$\partial\psi = \partial\Omega + \frac{4i}{1-\nu}\,\frac{d^2A}{d\zeta^2}$$

and substitute this into the conjugate of (d.1), which then reads

$$\alpha - i\beta = \bar{A} + \bar{\zeta}\,\frac{dA}{d\zeta} + \frac{dB}{d\zeta} + \frac{8h^2}{3(1-\nu)n}\,\frac{d^2A}{d\zeta^2} - \frac{2ih^2}{3n}\,\partial\Omega. \qquad (d.4)$$

By applying the Laplacian to equation (n.2) we get

$$\nabla^2\psi = \nabla^2\Omega$$

so that equation (f) is no more than a disguised form of equation (n.3) controlling the twist.

By allowing $n$ to tend to infinity in these results we suppress the deformations due to the shear forces. As may also be verified directly, this shows that in Kirchhoff's theory there can be no transverse twist (indeed, there is no transverse rotation).

By starting with the general expressions (d.3) and (d.4) for the displacements we obtain in succession for the moments

$$\frac{M_x + M_y}{D(1+\nu)} - i(1-\nu)(\psi-\Omega) = 4\,\frac{dA}{d\zeta}, \qquad \text{(m.1)}$$

an equation containing (r.2) as its imaginary part, and

$$\frac{M_x - M_y - 2iM_{xy}}{D(1-\nu)} = \partial(\alpha - i\beta) = 2\bar{\zeta}\,\frac{d^2A}{d\zeta^2} + 2\,\frac{d^2B}{d\zeta^2} + \frac{16h^2}{3(1-\nu)n}\,\frac{d^3A}{d\zeta^3}$$

$$- \frac{2ih^2}{3n}\,\partial\partial\Omega\,, \qquad \text{(m.2)}$$

$$\frac{T_x - iT_y}{D(1-\nu)} = -i\partial\psi = -i\partial\Omega + \frac{4}{1-\nu}\,\frac{d^2A}{d\zeta^2}. \qquad \text{(m.3)}$$

## 8.10. The Edge Effect

The structure just obtained for the solution of the equations for the bending of plates of moderate thickness separates the part depending on the biharmonic operator from that depending on the elliptic operator controlling the transverse twist.

It is necessary to observe that this separation does not coincide with a simple superposition of a Kirchhoff solution and a solution based only on the transverse twist, for in allowing $n$ to tend to infinity we obtain $\Omega = 0$ from (n.3), and we suppress in the rotations and moments the other analytic terms connected with the deformations due to shear forces. The simple superposition does apply, however, to the transverse displacement and shear force fields.

The change of analytic function

$$B = C - \frac{8h^2}{3(1-\nu)n}\,\frac{dA}{d\zeta}$$

would be enough to bring the general solution into the form in which the superposition of a Kirchhoff solution and a transverse twist applies now to the rotations, moments, and shear forces, and further to the transverse displacement, which becomes

$$W = -\mathrm{Re}\{\bar{\zeta}A + C - \frac{8h^2}{3(1-\nu)n}\frac{dA}{d\zeta}\}.$$

The boundary terms arising from the transverse twist affect all quantities except $W$, although it is impossible to separate the problem governing the transverse twist from the general problem:

$$\alpha_n = -\frac{2h^2}{3n}\frac{\partial\Omega}{\partial s}, \qquad \alpha_t = \frac{2h^2}{3n}\frac{\partial\Omega}{\partial n},$$

$$M_n = -D(1-\nu)\frac{2h^2}{3n}\left(\frac{\partial}{\partial s}\frac{\partial\Omega}{\partial n} - \dot{\theta}\frac{\partial\Omega}{\partial s}\right),$$

$$M_{nt} = D(1-\nu)\frac{h^2}{3n}\left(\frac{\partial^2}{\partial n^2} - \frac{\partial^2}{\partial s^2} - \dot{\theta}\frac{\partial}{\partial n}\right)\Omega$$

$$= D(1-\nu)\left(-\Omega + \frac{2h^2}{3n}\frac{\partial^2\Omega}{\partial n^2}\right) = D(1-\nu)\left[\Omega - \frac{2h^2}{3n}\left(\frac{\partial^2}{\partial s^2} + \dot{\theta}\frac{\partial}{\partial n}\right)\Omega\right],$$

$$T_n = -D(1-\nu)\frac{\partial\Omega}{\partial s}.$$

The general solution of Kirchhoff's problem is obtained by setting $\Omega = 0$ and making $n$ infinite. It is appropriate here to calculate Kirchhoff's shear forces

$$K_x - iK_y = T_x - iT_y + i\partial M_{xy}.$$

Now we have

$$\frac{-M_{xy}}{D(1-\nu)} = \mathrm{Im}\left(\bar{\zeta}\frac{d^2A}{d\zeta^2} + \frac{d^2B}{d\zeta^2}\right) = \frac{1}{2i}\left(\bar{\zeta}\frac{d^2A}{d\zeta^2} + \frac{d^2B}{d\zeta^2} - \zeta\overline{\frac{d^2A}{d\zeta^2}} - \overline{\frac{d^2B}{d\zeta^2}}\right),$$

and the application of the rule for differentiating by the operator $\partial$ yields

$$\frac{K_x - iK_y}{D(1-\nu)} = \frac{4}{1-\nu}\frac{d^2A}{d\zeta^2} - \bar{\zeta}\frac{d^3A}{d\zeta^3} + \overline{\frac{d^2A}{d\zeta^2}} - \frac{d^3B}{d\zeta^3}.$$

## 8.11. Torsion of a Plate

The representation developed in the last section permits the systematic formulation of solutions for the equations of the theory of moderately thick plates, based on a choice of the functions A and B of a complex variable and the choice of a solution for equation (n.3). From each solution one may determine which boundary conditions would apply for any shape of plate. The problem is then to obtain, eventually by superposition, the boundary conditions actually desired.

The examples to be used in this and the next section to illustrate the theory have the advantage of simplicity and allow comparison with the results of other theories. For example, if a plate is rectangular with no applied stresses on the edges $y = \pm b$ and no transverse load, then it may also be treated as a bar in bending or torsion. A solution of the plate-theory type may therefore be compared with a solution of Saint-Venant's type.

We look first at the problem of torsion (Figure 8.4),

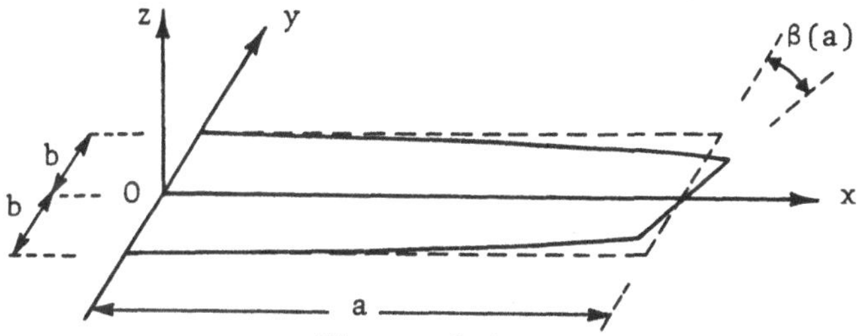

Figure 8.4

with the corresponding boundary conditions

$$M_x = 0, \quad \beta = 0, \quad w = 0 \quad \text{at} \quad x = 0,$$

$$M_x = 0, \quad \beta = \beta(a), \quad w = -y\beta(a) \quad \text{at} \quad x = a.$$

In Saint-Venant's torsion, the stress $\sigma_x$ in the longitudinal fibers vanishes and one would therefore expect to have $M_x \equiv 0$. That each cross section rotates about the x-axis leads us to anticipate a solution

$$\beta = x\theta, \quad W = -xy\theta,$$

with $\theta = \beta(a)/a$. We also know that among longitudinal fibers there are only tangential effects, represented here by $\tau_{xy}$ and $\tau_{xz}$, or rather by their resultants $M_{xy}$ and $T_z$, and that Saint-Venant's postulates $\tau_{yz} \equiv 0$, $\sigma_y \equiv 0$ respectively imply that $T_y \equiv 0$ and $M_y \equiv 0$. The lateral edges $y = \pm b$ are thus free of stresses if only we require that $M_{xy} = 0$ at $y = \pm b$.

The equilibrium equations (e.1) and (e.2) now reduce to

$$\partial M_{xy}/\partial y = T_x, \quad \partial M_{xy}/\partial x = 0, \quad \partial T_x/\partial x = 0,$$

and we see that $T_x$ and $M_{xy}$ depend only on $y$. The moment-rotation relations (8.16) give

$$0 = D(\frac{\partial \alpha}{\partial x}) = 0,$$

$$0 = D\nu \frac{\partial \alpha}{\partial x} = 0,$$

$$M_{xy} = D \frac{1-\nu}{2} (\frac{\partial \alpha}{\partial y} + \theta),$$

and show that $\alpha$ and $M_{xy}$ vary only with $y$.

Now we observe that Kirchhoff's condition $\beta + \partial W/\partial y \equiv 0$, as well as the second of (c.2), are satisfied identically, while the first of (c.2) becomes

$$T_x = 2nGh(\alpha - y\theta).$$

We thus have

$$T_x = \frac{dM_{xy}}{dy} = D \frac{1-\nu}{2} \frac{d^2\alpha}{dy^2} = 2nGh(\alpha - y\theta)$$

or, by rearrangement, the differential equation

$$\frac{d^2\alpha}{dy^2} - \frac{3n}{h^2} \alpha = - \frac{3n}{h^2} y\theta,$$

whose solution of appropriate parity in $y$ (i.e., odd) is

$$\alpha = \theta y + C \sinh (\sqrt{3n}\, \frac{y}{h}).$$

We see that

$$M_{xy} = D \frac{1-\nu}{2} [2\theta + \frac{\sqrt{3n}}{h} C \cosh (\sqrt{3n}\, \frac{y}{h})],$$

and may determine the constant $C$ from the condition that this twisting moment vanish for $y = \pm b$. Finally, then, we have

$$M_{xy} = D(1-\nu) \left[1 - \frac{\cosh (\sqrt{3n}\, y/h)}{\cosh (\sqrt{3n}\, b/h)}\right]\theta ,$$

$$T_x = -D(1-\nu) \frac{\sqrt{3n}}{h} \frac{\sinh (\sqrt{3n}\, y/h)}{\cosh (\sqrt{3n}\, b/h)} \theta ,$$

$$\alpha = \left[y - \frac{2h}{\sqrt{3n}} \frac{\sinh (\sqrt{3n}\, y/h)}{\cosh (\sqrt{3n}\, b/h)}\right]\theta,$$

the last result portraying in fact the warping of a cross section by torsion. We observe that if $b >> h$, then the transverse twist of the fibers

$$\Omega = \frac{1}{2} (\frac{\partial\beta}{\partial x} - \frac{\partial\alpha}{\partial y}) = \frac{\cosh (\sqrt{3n}\, y/h)}{\cosh (\sqrt{3n}\, b/h)} \theta$$

is well concentrated near the edges $y = \pm b$, and at the edges it reaches precisely the same value as the longitudinal twist.

According to our basic theory of moderately thick plates, the shear stresses in a section $x = \text{const.}$ are given by

$$\tau_{xz} = \frac{h^2-z^2}{2} \frac{3}{2h^3} T_x, \quad \tau_{xy} = \frac{3}{2h^3} z M_{xy}.$$

One may directly verify that these can be derived from the stress function

$$\Theta = (h^2-z^2)\left[\frac{\cosh(\sqrt{3n}\ y/h)}{\cosh(\sqrt{3n}\ b/h)} - 1\right], \qquad n = \frac{5}{6},$$

by the relations

$$\tau_{xz} = -G\theta\frac{\partial\Theta}{\partial y}, \qquad \tau_{xy} = G\theta\frac{\partial\Theta}{\partial z},$$

and this result can be compared with Saint-Venant's.

The same stress function results from the minimization of the integral

$$\iint\{(\frac{\partial\Theta}{\partial y})^2 + (\frac{\partial\Theta}{\partial z})^2 + 4\Theta\}\ dydz$$

over functions of the form

$$\Theta = (h^2-z^2)g(y) \qquad \text{with} \qquad g(\pm b) = 0.$$

On the other hand, it is rather delicate to justify this function as an approximation for $h << b$ to the exact solution found in Section 6.18.E for the torsion of a bar of rectangular cross-section.

The torsion constant for the plate, which is involved in the stiffness relation $C = GJ\theta$, can be calculated either by using the classical relation

$$J = 2\iint\Theta\ dydz,$$

or by direct calculation of the couple

$$C = \int_{-b}^{b}(M_{xy} - yT_x)dy.$$

Either approach yields

$$J = \frac{1}{3}(2b)(2h)^3\left(1 - \frac{\tanh\mu}{\mu}\right) \quad \text{with} \quad \mu = \sqrt{3n}\,\frac{b}{h}.$$

Kirchhoff's solution for the same problem is given by

$$\alpha = \theta y, \quad \beta = \theta x, \quad W = -\theta xy, \quad M_{xy} = D(1-\nu)\theta, \quad M_x = M_y = T_x = T_y \equiv 0.$$

In this instance one may obtain Kirchhoff's solution from that of moderately thick plates simply by suppressing the effect of transverse twist in the latter, i.e., by ignoring the terms involving ratios of hyperbolic functions. The boundary condition at $y = \pm b$ is no longer satisfied except in Kirchhoff's sense:

$$K_y = T_y - \frac{dM_{xy}}{dy} = 0.$$

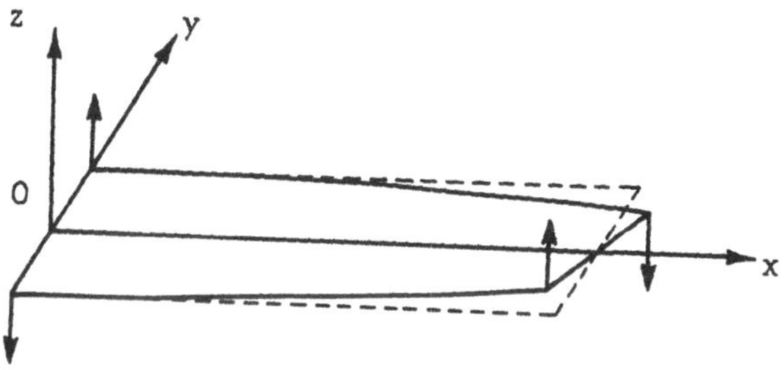


Figure 8.5

Therefore, by Kelvin and Tait's interpretation, the deformation reduces to the effect of four corner forces of intensity $2D(1-\nu)\theta$, and the torsion constant is slightly reduced by Kirchhoff's second constraint $\alpha + \partial W/\partial x = 0$ to the classical approximation $J = (2b)(2h)^3/3$.

## 8.12. Saint-Venant's Bending of a Plate

To study the bending of a plate by terminal moments and shear forces we again invoke Saint-Venant's hypotheses $\sigma_y \equiv 0$, $\tau_{xy} \equiv 0$ to infer as before that $M_y \equiv 0$ and $T_y \equiv 0$. With $q = 0$, the equilibrium equations (e.1) and (e.2) show that $T_x$ and $M_{xy}$ depend only on $y$, and that

$$\partial M_x/\partial x = T_x - dM_{xy}/dy. \qquad \text{(f.1)}$$

Equations (c.2) reduce here to

$$\beta + \frac{\partial W}{\partial y} = 0, \tag{f.2}$$

$$\alpha + \frac{\partial W}{\partial x} = \frac{1}{2nGh} T_x. \tag{f.3}$$

Eliminating W yields

$$\frac{\partial \beta}{\partial x} - \frac{\partial \alpha}{\partial y} = 2\Omega = - \frac{1}{2nGh} \frac{dT_x}{dy}, \tag{f.4}$$

and this shows that the transverse twist is a function only of $y$.

Equations (c.1) reduce to

$$M_x = D(1-\nu^2) \frac{\partial \alpha}{\partial x}, \tag{f.5}$$

$$\frac{\partial \beta}{\partial y} = -\nu \frac{\partial \alpha}{\partial x}, \tag{f.6}$$

$$M_{xy} = D \frac{1-\nu}{2}\left(\frac{\partial \beta}{\partial x} + \frac{\partial \alpha}{\partial y}\right). \tag{f.7}$$

The last four equations yield

$$\frac{\partial \alpha}{\partial x} = \frac{M_x}{D(1-\nu^2)}, \quad \frac{\partial \alpha}{\partial y} = \frac{M_{xy}}{D(1-\nu)} + \frac{1}{4nGh} \frac{dT_x}{dy},$$

$$\frac{\partial \beta}{\partial x} = \frac{M_{xy}}{D(1-\nu)} - \frac{1}{4nGh} \frac{dT_x}{dy}, \quad \frac{\partial \beta}{\partial y} = -\nu \frac{M_x}{D(1-\nu^2)}.$$

Equations of the Beltrami-Michell type follow from the elimination of $\alpha$ and $\beta$. It follows from the first one, $\partial M_x/\partial y = 0$, that $M_x$ depends only on $x$. The other,

$$-\frac{\nu}{D(1-\nu^2)} \frac{dM_x}{dx} = \frac{1}{D(1-\nu)} \frac{dM_{xy}}{dy} - \frac{1}{4nGh} \frac{d^2T_x}{dy^2},$$

can be combined with (f.1) to eliminate the twisting moment $M_{xy}$. The result is

$$\frac{1}{1+\nu} \frac{dM_x}{dx} = T_x - \frac{h^2}{3n} \frac{d^2T_x}{dy^2}$$

and the two members, one depending only on $x$ and the other only on $y$, must equal a common constant $c$. Hence

$$M_x = (1+\nu)c(x-a),$$

the constant of integration being chosen so as to annul the bending at the terminal section $x = a$. Also,

$$T_x = c + A \cosh (\sqrt{3n}\, \tfrac{y}{h})$$

the second term reflecting the necessary parity in $y$ (i.e., even).

Equation (f.1) may now be solved for $M_{xy}$ to yield

$$M_{xy} = -\nu cy + \frac{h}{\sqrt{3n}} A \sinh (\sqrt{3n}\, \tfrac{y}{h}),$$

the constant of integration again being fixed by the type of parity required. In order that this moment vanish at $y = \pm b$ it is necessary that

$$\frac{h}{\sqrt{3n}} A \sinh (\sqrt{3n}\, \tfrac{b}{h}) = \nu cb.$$

At the same time one may express the intensity of the solution as a function of the total vertical shear force

$$V_x = \int_{-b}^{b} T_x\, dy = 2bc + 2A \frac{h}{\sqrt{3n}} \sinh (\sqrt{3n}\, \tfrac{b}{h}) = (1+\nu)2cb.$$

The results are

$$T_x = \frac{V_x}{2(1+\nu)b}\left[1 + \nu\sqrt{3n}\, \frac{b}{h}\, \frac{\cosh (\sqrt{3n}\, y/h)}{\sinh (\sqrt{3n}\, b/h)}\right],$$

$$M_x = \frac{V_x}{2b}(x-a), \quad M_{xy} = -\nu \frac{V_x}{2(1+\nu)b}\left[y - b\, \frac{\sinh(\sqrt{3n}\, y/h)}{\sinh(\sqrt{3n}\, b/h)}\right],$$

$$\beta = -\frac{\nu}{D(1-\nu^2)} \frac{V_x}{2b} y(x-a) = \frac{-3\nu}{2Eh^3} \frac{V_x}{2b} y(x-a),$$

$$\alpha = \frac{3}{2Eh^3} \frac{V_x}{2b}\left\{\frac{x^2}{2} - ax + \nu\left[-\frac{y^2}{2} + \frac{2bh}{\sqrt{3n}}\, \frac{\cosh(\sqrt{3n}\, y/h)}{\sinh(\sqrt{3n}\, b/h)} + k\right]\right\},$$

where the constant $k$ can be determined by the approximate condition for clamping at the section $x = 0$:

$$\int_{-b}^{b} \alpha \, dy = 0 \Rightarrow k = \frac{1}{6} b^2 - \frac{2h^2}{3n} .$$

Finally, the transverse displacement is

$$W = \frac{3}{2Eh^3} \frac{V_x}{2b} \{- \frac{x^3}{6} + a \frac{x^2}{2} + (1+\nu) \frac{2h^2}{3n} x + \nu[\frac{y^2}{2} (x-a) - \frac{b^2}{6} x] + c\},$$

where the constant $c$, denoting a rigid-body vertical translation, may also be determined by an approximation condition of clamping at the section $x = 0$. Two notable features of the expression for $W$ are the transverse (anticlastic) curvature due to a Poisson's effect, and the term involving $n$ and representing the shear deformation.

The stresses may be expressed in terms of stress functions by

$$\tau_{xz} = \frac{3V_x}{4bh^3} (\frac{\partial \Phi}{\partial z} - \frac{\nu}{1+\nu} \frac{\partial \Psi}{\partial y}),$$

$$\tau_{xy} = \frac{3V_x}{4bh^3} (\frac{\partial \Phi}{\partial y} + \frac{\nu}{1+\nu} \frac{\partial \Psi}{\partial z}).$$

Here, the principal bending stress function $\Phi$ is given by

$$\Phi = \frac{1}{6} (-z^3 + 3h^2 z)$$

and corresponds precisely to that for a bar of rectangular section (cf. Section 6.18.H). The secondary stress function $\Psi$, representing Poisson's effect, is given by

$$\Psi = \frac{h^2 - z^2}{2} \left[ y - b \frac{\sinh (\sqrt{3n}\, y/h)}{\sinh (\sqrt{3n}\, b/h)} \right],$$

and is a variational approximation of the form $\Psi = (h^2 - z^2) g(y)$ minimizing the functional

$$\iint [(\frac{\partial \Psi}{\partial y})^2 + (\frac{\partial \Psi}{\partial z})^2 - 2y\Psi] \, dydz$$

among functions vanishing on the edges $y = \pm b$.

By letting $n$ tend to infinity we find Kirchhoff's solution, which suppresses not only the edge effect but also the shear deformation. The twisting moment $M_{xy}$ no longer vanishes on the edges $y = \pm b$, but Kirchhoff's shear force $K_y$ does. The non-vanishing boundary loads are a uniform distribution of Kirchhoff's shear force

$$K_x = T_x + \frac{\partial M_{xy}}{\partial y} = \frac{V_x}{2b} \frac{1-\nu}{1+\nu}$$

at $x = 0,a$; additional corner forces of intensity $\frac{\nu}{1+\nu} V_x$; and a uniformly distributed bending moment $M_x = -aV_x/2b$ at $x = 0$ (Figure 8.6).

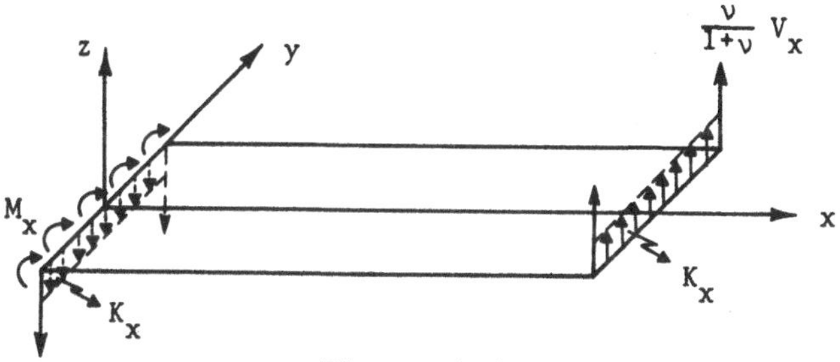


Figure 8.6

## 8.13. Particular Solutions for Transverse Load

If one has a particular solution of the plate bending equations under a given transverse load $q(x,y)$, but without regard for the boundary conditions, then the problem with given boundary conditions can be reduced to that of a complementary solution with vanishing transverse load and modified boundary conditions. With the index zero indicating the particular solution, the complementary solution should satisfy

$$\alpha_n = \bar{\alpha}_n - \alpha_{on} \quad \text{or} \quad M_n = \bar{M}_n - M_{on},$$

$$\alpha_t = \bar{\alpha}_t - \alpha_{ot} \quad \text{or} \quad M_{nt} = \bar{M}_{nt} - M_{ont},$$

$$W = \bar{W} - W_o \quad \text{or} \quad T_n = \bar{T}_n - T_{on},$$

for the alternatives of the theory of moderately thick plates, and

$$\partial W/\partial n = \partial\bar{W}/\partial n - \partial W_o/\partial n \quad \text{or} \quad M_n = \bar{M}_n - M_{on},$$

$$W = \bar{W} - W_o \quad \text{or} \quad \begin{cases} T_n + \partial M_{nt}/\partial s = \bar{K}_n - T_{on} - \partial M_{ont}/\partial s \\ \Delta_i M_{nt} = \bar{Z}_i - \Delta_i M_{nto}, \end{cases}$$

for Kirchhoff's theory.

When the load distribution is a polynomial, a particular solution may be constructed by finding functions $(\phi, \lambda)$ such that

$$\nabla^2\phi = q(x,y),$$

$$\nabla^2\lambda = (1-\nu)\phi.$$

Then the displacements

$$\alpha = -\frac{1}{D(1-\nu)}\frac{\partial\lambda}{\partial x}, \qquad \beta = -\frac{1}{D(1-\nu)}\frac{\partial\lambda}{\partial y},$$

$$W = -\frac{1}{2nGh}\phi + \frac{1}{D(1-\nu)}\lambda,$$

and the moments and shear forces

$$M_x = -\phi + \frac{\partial^2\lambda}{\partial y^2}, \quad M_y = -\phi + \frac{\partial^2\lambda}{\partial x^2}, \quad M_{xy} = -\frac{\partial^2\lambda}{\partial x\partial y},$$

$$T_x = -\frac{\partial\phi}{\partial x}, \quad T_y = -\frac{\partial\phi}{\partial y},$$

satisfy the general equations of the theory of moderately thick plates. In terms of $\phi$ and $\lambda$, the boundary contributions are

$$\alpha_{on} = -\frac{1}{D(1-\nu)}\frac{\partial\lambda}{\partial n}, \qquad \alpha_{ot} = -\frac{1}{D(1-\nu)}\frac{\partial\lambda}{\partial s},$$

$$W_o = -\frac{1}{2nGh}\phi + \frac{1}{D(1-\nu)}\lambda,$$

$$M_{on} = -\frac{1+\nu}{2}\phi + \frac{1}{2}\left(\frac{\partial^2\lambda}{\partial s^2} - \frac{\partial^2\lambda}{\partial n^2} + \dot{\theta}\frac{\partial\lambda}{\partial n}\right) = -\phi + \frac{\partial^2\lambda}{\partial s^2} + \dot{\theta}\frac{\partial\lambda}{\partial n},$$

$$M_{ont} = -\frac{\partial}{\partial s}\frac{\partial\lambda}{\partial n} + \frac{1}{2}\dot{\theta}\frac{\partial\lambda}{\partial s}, \qquad T_{on} = -\frac{\partial\phi}{\partial n}.$$

We note that in this particular solution the transverse twist is zero. By letting $n$ approach infinity we find the corresponding particular solution for Kirchhoff's theory.

## 8.14. Solutions in Polar Coordinates

By starting with tensor transformations

$$\alpha_r + i\alpha_\theta = e^{-i\theta}(\alpha + i\beta),$$

$$M_r + M_\theta = M_x + M_y,$$

$$M_r - M_\theta - 2iM_{r\theta} = e^{2i\theta}(M_x - M_y - 2iM_{xy}),$$

$$T_r - iT_\theta = e^{i\theta}(T_x - iT_y),$$

$$\partial = \frac{\partial}{\partial x} - i\frac{\partial}{\partial y} = e^{-i\theta}\left(\frac{\partial}{\partial r} - i\frac{1}{r}\frac{\partial}{\partial\theta}\right),$$

analogous to those used earlier for the treatment of boundary conditions in the natural directions, the functional of the canonical variational principle assumes the form

$$\int\Big[M_r\frac{\partial\alpha_r}{\partial r} + M_\theta\left(\frac{1}{r}\frac{\partial\alpha_\theta}{\partial\theta} + \frac{\alpha_r}{r}\right) + M_{r\theta}\left(\frac{\partial\alpha_\theta}{\partial r} - \frac{\alpha_\theta}{r} + \frac{1}{r}\frac{\partial\alpha_r}{\partial\theta}\right)$$

$$+ T_r\left(\alpha_r + \frac{\partial W}{\partial r}\right) + T_\theta\left(\alpha_\theta + \frac{1}{r}\frac{\partial W}{\partial\theta}\right)\Big]\, r\,drd\theta$$

$$- \frac{1}{2D(1-\nu^2)}\int[(M_r+M_\theta)^2 + 2(1+\nu)(M_{r\theta}^2 - M_rM_\theta)]\, r\,drd\theta$$

$$- \frac{1}{4nGh}\int(T_r^2 + T_\theta^2)r\,drd\theta - \int qW\, r\,drd\theta$$

$$- \oint(\overline{M}_r\alpha_r + \overline{M}_{r\theta}\alpha_\theta + \overline{T}_rW)r\,d\theta.$$

The variational equations connected with the displacements $\alpha_r$, $\alpha_\theta$, and $W$ are the equilibrium equation

$$\frac{\partial}{\partial r}(rM_r) - M_\theta + \frac{\partial}{\partial\theta} M_{r\theta} = rT_r,$$

$$\frac{\partial}{\partial r}(rM_{r\theta}) + M_{r\theta} + \frac{\partial}{\partial\theta} M_\theta = rT_\theta,$$

$$\frac{\partial}{\partial r}(rT_r) + \frac{\partial}{\partial\theta} T_\theta = -rq.$$

The moment-rotation and shear deformation equations follow from variations of the bending moments and shear forces:

$$M_r = D(\frac{\partial\alpha_r}{\partial r} + \nu\,\frac{1}{r}\,\frac{\partial\alpha_\theta}{\partial\theta} + \nu\,\frac{\alpha_r}{r}), \quad M_\theta = D(\frac{1}{r}\,\frac{\partial\alpha_\theta}{\partial\theta} + \frac{\alpha_r}{r} + \nu\,\frac{\partial\alpha_r}{\partial r}),$$

$$M_{r\theta} = D\,\frac{1-\nu}{2}\,(\frac{\partial\alpha_\theta}{\partial r} - \frac{\alpha_\theta}{r} + \frac{1}{r}\,\frac{\partial\alpha_r}{\partial\theta}),$$

$$T_r = 2nGh(\alpha_r + \frac{\partial W}{\partial r}), \quad T_\theta = 2nGh(\alpha_\theta + \frac{1}{r}\,\frac{\partial W}{\partial\theta}).$$

The general solution of these equations may be deduced at once from the general solution in Cartesian coordinates, and is

$$W = -\mathrm{Re}\{\bar{\zeta}A + B\},$$

$$\alpha_r - i\alpha_\theta = e^{i\theta}(\bar{A} + \bar{\zeta}\,\frac{dA}{d\zeta} + \frac{dB}{d\zeta} + \frac{8h^2}{3(1-\nu)n}\,\frac{d^2A}{d\zeta^2}) - \frac{2ih^2}{3n}(\frac{\partial}{\partial r} - \frac{i}{r}\,\frac{\partial}{\partial\theta})\Omega,$$

$$\frac{M_r + M_\theta}{D(1+\nu)} - i(1-\nu)(\psi-\Omega) = 4\,\frac{dA}{d\zeta},$$

$$\frac{M_r - M_\theta - 2iM_{r\theta}}{D(1-\nu)} = e^{2i\theta}(2\bar{\zeta}\,\frac{d^2A}{d\zeta^2} + 2\,\frac{d^2B}{d\zeta^2} + \frac{16h^2}{3(1-\nu)n}\,\frac{d^3A}{d\zeta^3})$$

$$- \frac{2h^2}{3n}\left[\frac{2}{r}\,\frac{\partial^2\Omega}{\partial\theta\partial r} + i(\frac{\partial^2\Omega}{\partial r^2} + \frac{1}{r}\,\frac{\partial\Omega}{\partial r} - \frac{1}{r^2}\,\frac{\partial^2\Omega}{\partial\theta^2})\right],$$

$$\frac{T_r - iT_\theta}{D(1-\nu)} = e^{i\theta}\,\frac{4}{1-\nu}\,\frac{d^2A}{d\zeta^2} - (\frac{1}{r}\,\frac{\partial}{\partial\theta} + i\,\frac{\partial}{\partial r})\Omega.$$

## 8.15. Axisymmetric Bending

Consider the particular case of solutions independent of the polar angle $\theta$. By symmetry, $T_\theta = M_{r\theta} = \alpha_\theta = 0$, and the equations reduce to

$$\frac{d}{dr}(rM_r) - M_\theta = rT_r, \quad \frac{d}{dr}(rT_r) = -rq(r),$$

$$M_r = D\left(\frac{d\alpha_r}{dr} + \nu\frac{\alpha_r}{r}\right), \quad M_\theta = D\left(\frac{\alpha_r}{r} + \nu\frac{d\alpha_r}{dr}\right),$$

$$T_r = 2nGh\left(\alpha_r + \frac{dW}{dr}\right).$$

The edge effect is absent, because due to the symmetry we have

$$2\Omega = \frac{\partial\alpha_\theta}{\partial r} - \frac{1}{r}\frac{\partial\alpha_r}{\partial\theta} + \frac{\alpha_\theta}{r} = 0.$$

In the absence of transverse load, elementary integration yields a solution with four arbitrary constants $a$, $b$, $c$ $e$ and a parameter $R$ to render the argument of the logarithmic function dimensionless:

$$\frac{M_r}{D} = 2a(1+\nu)\ell n\frac{r}{R} + (1-\nu)a + (1+\nu)b - (1-\nu)\frac{c}{r^2},$$

$$\frac{M_\theta}{D} = 2a(1+\nu)\ell n\frac{r}{R} - (1-\nu)a + (1+\nu)b + (1-\nu)\frac{c}{r^2},$$

$$\frac{T_r}{D} = \frac{4a}{r}, \quad \alpha_r = 2ar\,\ell n\frac{r}{R} - ar + br + \frac{c}{r},$$

$$W = \frac{8ah^2}{3(1-\nu)n}\ell n\frac{r}{R} - ar^2\,\ell n\frac{r}{R} + ar^2 - \frac{b}{2}r^2 - c\,\ell n\frac{r}{R} + e.$$

For an annular plate simply supported at $r = R$ and loaded at $r = \rho < R$ by a shear force $T_r = T$, the boundary conditions are $W = M_r = 0$ at $r = R$, $T_r = T$ and $M_r = 0$ at $r = \rho$. The values of the constants are found to be

$$a = \frac{T\rho}{4D}, \qquad e = \frac{b}{2} R^2 - aR^2,$$

$$(1+\nu)b = -(1-\nu)a - \frac{\rho^2}{R^2-\rho^2} 2a(1+\nu)\ln\frac{R}{\rho},$$

$$(1-\nu)c = -\frac{\rho^2 R^2}{R^2-\rho^2} 2a(1+\nu)\ln\frac{R}{\rho}.$$

If measured positively in the direction of $Oz$, the total transverse force applied along the interior edge is $Q = -2\pi T\rho$.

In the limiting case where $\rho \to 0$ and $T \to \infty$ with $T\rho = -Q/2\pi = \text{const.}$, this solution does not correspond, as one might expect, to the problem of the complete plate simply supported at $r = R$ and loaded at its center. In that limit we find $M_r = 0$ and $\alpha_r \to \infty$ at $r = 0$, while for a centrally loaded plate one would want $\alpha_r = 0$ with no restriction on $M_r$.

For an annular plate simply supported at $r = R$ and clamped to a piston at $r = \rho$, the boundary conditions are $W = M_r = 0$ at $r = R$, and $T_r = T$, $\alpha_r = 0$ at $r = \rho$. The constants are found to be

$$a = \frac{T\rho}{4D} = -\frac{Q}{8\pi D}, \qquad e = \frac{b}{2} R^2 - aR^2,$$

$$\left(\frac{1+\nu}{1-\nu} + \frac{\rho^2}{R^2}\right)\frac{b}{a} = \frac{\rho^2}{R^2}\left(1 + 2\ln\frac{R}{\rho}\right) - 1,$$

$$\left(\frac{1+\nu}{1-\nu} + \frac{\rho^2}{R^2}\right)\frac{c}{a} = \frac{2\rho^2}{1-\nu}\left[1 + (1+\nu)\ln\frac{R}{\rho}\right].$$

Here the passage to the limit $\rho = 0$ yields $b/a = -(1-\nu)/(1+\nu)$, $c/a = 0$, and

$$\frac{8\pi D}{Q} W = \left[r^2 - \frac{8h^2}{3(1-\nu)n}\right]\ln\frac{r}{R} + \frac{3+\nu}{2(1+\nu)}(R^2-r^2),$$

$$\frac{4\pi D}{Q}\alpha_r = -r\ln\frac{r}{R} + \frac{r}{1+\nu}, \qquad T_r = -\frac{Q}{2\pi}\frac{1}{r},$$

$$M_r = -\frac{Q}{4\pi}(1+\nu)\ln\frac{r}{R}, \quad M_\theta = -\frac{Q}{4\pi}[(1+\nu)\ln\frac{r}{R} - 1 + \nu].$$

The shear force and bending moments exhibit singularities at the origin, logarithmic for the latter and like $r^{-1}$ for the former. The displacement is also logarithmically singular, a result permitted by the shear deformations. If $n$ tends to infinity we recover Kirchhoff's solution, in which the singularities disappear and the central deflection takes the value

$$\frac{Q}{8\pi D}\,\frac{3+\nu}{2(1+\nu)}\,R^2.$$

The analytic functions $(A,B)$ for the foregoing problem are

$$A = \frac{Q}{8\pi D}\left[-\zeta\left(\ln\frac{\zeta}{R} - 1\right) + \frac{1-\nu}{2(1+\nu)}\,\zeta\right],$$

$$B = \frac{Q}{8\pi D}\left[\frac{8h^2}{3(1-\nu)n}\,\ln\frac{\zeta}{R} - \frac{3+\nu}{2(1+\nu)}\,R^2\right].$$

They both contain logarithimic singularities and only that of $B$ results from the shear deformations. On the other hand, the auxiliary function

$$C = B + \frac{8h^2}{3(1-\nu)n}\,\frac{dA}{n\zeta}$$

introduced in Section 8.10 remains regular.

# Bibliography

E. Almansi, Un teorema sulle deformazioni elastiche dei solidi isotropi. Atti della Reale Academia dei Lincei (Roma), Ser. 5, Rendiconti **16**, 1° Semestre, 865-867 (1907).

________, L'ordinaria teoria dell' elasticità e la teoria delle deformazione finite. *Op. cit.* **26**, 2° Semestre, 3-8 (1917).

M. A. Biot, A hydrodynamic analogy for shearing stress distribution in bending. Journal of Applied Physics **9**, 39-43 (1938).

P. Cicala, Il centro di taglio nei solidi cilindrici. Atti della Reale Academia delle Scienze de Torino **70** (3), 356-371 (1935).

R. Courant and D. Hilbert, *Methods of Mathematical Physics*, Vol. I. Interscience, New York, 1953, pp. 268-272.

B. Fraeijs de Veubeke, Calcul des cadres bordant une ouverture circulaire dans un champ plan de tensions. Mécanique, 240-248 (1950).

________, La diffusion des inconnues hyperstatiques. Bulletin du Service Technique de l'Aéronautique, Mémoire **24**, Bruxelles (1951).

________, Aspects cinématiques et énergétiques de la flexion sans torsion. Academie Royale de Belgique, Classe de Sciences, Mémoires, Collection in-8°, **29** (1955).

________, Flexion et extension de plaques d'épaisseur modérée. *Op. cit.* **31** (1959).

________, A new variational principle for finite elastic deformations. International Journal of Engineering Sciences **10**, 745-763 (1972).

B. Fraeijs de Veubeke and G. Sander, An equilibrium model for plate bending. International Journal of Solids and Structures **4**, 447-468 (1968).

A. E. Green, On Reissner's theory of bending of elastic plates. Quarterly of Applied Mathematics **7**, 223-228 (1949).

A. E. Green and J. E. Adkins, *Large Elastic Deformations and Non-linear Continuum Mechanics*. Clarendon Press, Oxford, 1960.

J. Hadamard, *Lecons sur la Propagation des Ondes et les Équations de l'Hydrodynamique*. Hermann, Paris, 1903.

H. Jeffreys, *Cartesian Tensors*. The University Press, Cambridge, 1931.

G. Kirchhoff, Ueber das Gleichgewicht und die Bewegung einer elastichen Scheibe. Journal für die Reine und Angewandte Mathematik **40**, 51-88 (1850).

A. E. H. Love, *A Treatise on the Mathematical Theory of Elasticity*, Fourth Edition. The University Press, Cambridge, 1927.

J. H. Michell, On the direct determination of stress in an elastic solid, with application to the theory of plates. Proceedings of the London Mathematical Society **31**, 100-124 (1899).

N. I. Muskhelishvili, *Some Basic Problems of the Mathematical Theory of Elasticity*, Third Edition, translated from the Russian. Noordhoff, Gronigen, 1953.

G. L. Neidhardt and E. Sternberg, On the transmission of a concentrated load into the interior of an elastic body. Journal of Applied Mechanics **23**, 541-554 (1956).

C. E. Pearson, *Theoretical Elasticity*. Harvard University Press, Cambridge (USA), 1959.

J. Piola, La mecanica dé corpi naturalmente estesi, trattato col calcolo delle variazioni. Opuscoli Matematici e Fisici di Diversi Autori (Milano) **1**, 201-236 (1832).

E. Reissner, On the theory of bending of elastic plates. Journal of Mathematics and Physics **23**, 184-191 (1944).

________, On bending of elastic plates. Quarterly of Applied Mathematics **5**, 55-68 (1947).

________, On variational principles in elasticity, in *Proceedings of the Eighth Symposium in Applied Mathematics*. American Mathematical Society, New York, 1958.

I. Sokolnikoff, *Mathematical Theory of Elasticity*. McGraw-Hill, New York, 1946.

A. C. Stevenson, On the equilibrium of plates. Philosophical Magazine **33** (7), 639-661 (1942).

E. Sternberg and J. K. Knowles, Minimum energy characterizations of Saint-Venant's solution to the relaxed Saint-Venant's problem. Archive of Rational Mechanics and Analysis **21**, 89-107 (1966).

S. Timoshenko and J. N. Goodier, *Theory of Elasticity*. McGraw-Hill, New York, 1951.

A. J. C. Barré de Saint-Venant, Memoire sur la torsion des prismes. Mémoires par Divers Savants de l'Academie des Science (Paris) **14**, p. 233 (1855).

________, Memoire sur la flexion des prismes élastiques... Journal de Mathématiques Pures et Appliquées (Liouville) **1**, p. 89 (1856).

E. Trefftz, Über den Schubmittelpunkt in einem durch eine Einzellast gebogenen Balken. Zeitschrift für Angewandte Mathematik und Mechanik **15**, 220-225 (1935).

V. Volterra, Sur l'équilibre de corps élastiques multiplement connexes. Annales Scientifiques de l'École Normale Supérieure (Paris), **24** (3), 401-517 (1907).

C. Weber, Übertragung des Drehmomentes in Balken mit doppelflanschigem Querschnitt. Zeitschrift für Angewandte Mathematik und Mechanik **6**, 85-97 (1926).

C. Weber and W. Gunther, *Torsionstheorie*. F. Vieweg, Braunschweig, 1958.

G. Weingarten, Sulle superficie di discontinuità nella teoria della elasticità dei corpi solidi. Atti della Reale Academia dei Lincei (Roma), Ser. 5, Rendiconti **10**, 57-60 (1901).

# Index

The variational equations connected with the displacements $\alpha_r$, $\alpha_\theta$, and $W$ are the equilibrium equation

$$\frac{\partial}{\partial r}(rM_r) - M_\theta + \frac{\partial}{\partial\theta} M_{r\theta} = rT_r,$$

$$\frac{\partial}{\partial r}(rM_{r\theta}) + M_{r\theta} + \frac{\partial}{\partial\theta} M_\theta = rT_\theta,$$

$$\frac{\partial}{\partial r}(rT_r) + \frac{\partial}{\partial\theta} T_\theta = -rq.$$

The moment-rotation and shear deformation equations follow from variations of the bending moments and shear forces:

$$M_r = D(\frac{\partial\alpha_r}{\partial r} + \nu\,\frac{1}{r}\,\frac{\partial\alpha_\theta}{\partial\theta} + \nu\,\frac{\alpha_r}{r}), \quad M_\theta = D(\frac{1}{r}\,\frac{\partial\alpha_\theta}{\partial\theta} + \frac{\alpha_r}{r} + \nu\,\frac{\partial\alpha_r}{\partial r}),$$

$$M_{r\theta} = D\,\frac{1-\nu}{2}\,(\frac{\partial\alpha_\theta}{\partial r} - \frac{\alpha_\theta}{r} + \frac{1}{r}\,\frac{\partial\alpha_r}{\partial\theta}),$$

$$T_r = 2nGh(\alpha_r + \frac{\partial W}{\partial r}), \quad T_\theta = 2nGh(\alpha_\theta + \frac{1}{r}\,\frac{\partial W}{\partial\theta}).$$

The general solution of these equations may be deduced at once from the general solution in Cartesian coordinates, and is

$$W = -\mathrm{Re}\{\bar{\zeta}A + B\},$$

$$\alpha_r - i\alpha_\theta = e^{i\theta}(\bar{A} + \bar{\zeta}\,\frac{dA}{d\zeta} + \frac{dB}{d\zeta} + \frac{8h^2}{3(1-\nu)n}\,\frac{d^2A}{d\zeta^2}) - \frac{2ih^2}{3n}(\frac{\partial}{\partial r} - \frac{i}{r}\,\frac{\partial}{\partial\theta})\Omega,$$

$$\frac{M_r + M_\theta}{D(1+\nu)} - i(1-\nu)(\psi-\Omega) = 4\,\frac{dA}{d\zeta},$$

$$\frac{M_r - M_\theta - 2iM_{r\theta}}{D(1-\nu)} = e^{2i\theta}(2\bar{\zeta}\,\frac{d^2A}{d\zeta^2} + 2\,\frac{d^2B}{d\zeta^2} + \frac{16h^2}{3(1-\nu)n}\,\frac{d^3A}{d\zeta^3})$$

$$- \frac{2h^2}{3n}\left[\frac{2}{r}\,\frac{\partial^2\Omega}{\partial\theta\partial r} + i(\frac{\partial^2\Omega}{\partial r^2} + \frac{1}{r}\,\frac{\partial\Omega}{\partial r} - \frac{1}{r^2}\,\frac{\partial^2\Omega}{\partial\theta^2})\right],$$

$$\frac{T_r - iT_\theta}{D(1-\nu)} = e^{i\theta}\,\frac{4}{1-\nu}\,\frac{d^2A}{d\zeta^2} - (\frac{1}{r}\,\frac{\partial}{\partial\theta} + i\,\frac{\partial}{\partial r})\Omega.$$

# Applied Mathematical Sciences



Vol. 1 F. John
**Partial Differential Equations**
Third edition
ISBN 0-387-90327-5

Vol. 2 L. Sirovich
**Techniques of Asymptotic Analysis**
ISBN 0-387-90022-5

Vol. 3 J. Hale
**Theory of Functional Differential Equations**
ISBN 0-387-90203-1 (cloth)

Vol. 4 J. K. Percus
**Combinational Methods**
ISBN 0-387-90027-6

Vol. 5 R. von Mises and K. O. Friedrichs
**Fluid Dynamics**
ISBN 0-387-90028-4

Vol. 6 W. Freiberger and U. Grenander
**A Short Course in Computational Probability and Statistics**
ISBN 0-387-90029-2

Vol. 7 A. C. Pipkin
**Lectures on Viscoelasticity Theory**
ISBN 0-387-90030-6

Vol. 8 G. E. O. Giacaglia
**Perturbation Methods in Non-Linear Systems**
ISBN 0-387-90054-3

Vol. 9 K. O. Friedrichs
**Spectral Theory of Operators in Hilbert Space**
ISBN 0-387-90076-4

Vol. 10 A. H. Stroud
**Numerical Quadrature and Solution of Ordinary Differential Equations**
ISBN 0-387-90100-0

Vol. 11 W. A. Wolovich
**Linear Multivariable Systems**
ISBN 0-387-90101-9

Vol. 12 L. D. Berkovitz
**Optimal Control Theory**
ISBN 0-387-90106-X

Vol. 13 G. W. Bluman and J. D. Cole
**Similarity Methods for Differential Equations**
ISBN 0-387-90107-8

Vol. 14 T. Yoshizawa
**Stability Theory and the Existence of Periodic Solutions and Almost Periodic Solutions**
ISBN 0-387-90112-4

Vol. 15 M. Braun
**Differential Equations and Their Applications**
ISBN 0-387-90114-0

Vol. 16 S. Lefschetz
**Applications of Algebraic Topology**
ISBN 0-387-90137-X

Vol. 17 L. Collatz and W. Wetterling
**Optimization Problems**
ISBN 0-387-90143-4

Vol. 18 U. Grenander
**Pattern Synthesis**
**Lectures in Pattern Theory Vol. I**
ISBN 0-387-90174-4

Vol. 19 J. E. Marsden and M. McCracken
**The Hopf Bifurcation and Its Applications**
ISBN 0-387-90200-7

Vol. 20 R. D. Driver
**Ordinary and Delay Differential Equations**
ISBN 0-387-90231-7

Vol. 21 R. Courant and K. O. Friedrichs
**Supersonic Flow and Shock Waves**
ISBN 0-387-90232-5 (cloth)

Vol. 22 N. Rouche, P. Habets, and M. Laloy
**Stability Theory by Liapunov's Direct Method**
ISBN 0-387-90258-9

Vol. 23 J. Lamperti
**Stochastic Processes**
**A Survey of the Mathematical Theory**
ISBN 0-387-90275-9

Vol. 24 U. Grenander
**Pattern Analysis**
**Lectures in Pattern Theory Vol. II**
ISBN 0-387-90310-0

Vol. 25 B. Davies
**Integral Transforms and Their Applications**
ISBN 0-387-90313-5

Vol. 26 H. J. Kushner and D. S. Clark
**Stochastic Approximation Methods for Constrained and Unconstrained Systems**
ISBN 0-387-90341-0

Vol. 27 C. de Boor
**A Practical Guide to Splines**
ISBN 0-387-90356-9

Vol. 28 J. Keilson
**Markov Chain Models**
**Rarity and Exponentiality**
ISBN 0-387-90405-0